THE HOMEMAKER'S BOOK OF

Energy Savers

THE HOMEMAKER'S BOOK OF

Energy Savers

JEAN E. LAIRD

THE BOBBS-MERRILL CO., INC.

Indianapolis / New York

Published by The Bobbs-Merrill Company, Inc.
Indianapolis/New York
Manufactured in the United States of America
First paperback printing
Designed by: Susan Prindle

Library of Congress Cataloging in Publication Data

Laird, Jean E.
    The homemaker's book of energy savers that guarantee to save you money.
    1. Dwellings—Energy conservation. I. Title.
TJ163.5.D86L34   1982   644   82-1295
ISBN 0-672-52712-X   AACR2

Affectionately dedicated to:

John, Jayne, Joan-An,
Jerilyn & Jacquelyn

Contents

Introduction

As AMERICANS, we comprise 6 percent of the world's population and enjoy the world's highest standard of living. In doing so, we use one-third of the energy consumed on this planet. It is estimated that 30 percent or more of all energy used in the United States is now *wasted*. Did you know, for example, that if the hundreds of millions of dollars spent on the Alaskan oil pipeline had instead been invested in home insulation, the net energy savings would have been four times as great as the oil pumped from Alaskan oilfields? That's because, the experts say, we squander some 225 million barrels of oil in our homes each year due to poor insulation. Thus, we waste more energy than two-thirds of the world's population consumes. And, this waste is a luxury we can no longer afford.

Most of the energy we use in the United States comes from petroleum (crude oil)—about 16 million barrels a day (in 1980). Because domestic production falls short of our needs, more than 40 percent of the oil we consume has to be imported (in 1980). And our need for imports keeps increasing as energy demand rises.

In other words, if we continue using energy as we have become accustomed to do, experts contend we may run out of domestic oil supplies in the year 2007, and we may run out of natural gas even sooner! The severe winter of 1976–77 painfully dramatized the natural gas situation with its complex supply and economic problems.

Energy Saving in the Home

Let's take a look at facts closer to home. Do you realize that the home you live in—along with all the others in the country—consumes nearly one-fifth of all the energy produced in the United States? It's hard to believe, but it's true.

In past years, we have become so accustomed to low-cost energy that we have probably never considered it a commodity whose availability or cost must be reckoned in our day-to-day decisions. Today, just a look at our utility bills tells us all this has changed drastically. Energy in every form is becoming increasingly expensive. And, this year the price of electricity, natural gas, and heating oil, and the persistent economic inflation that they help create, are probably the main sources of consumer pain.

If you live in Southern California or on the East Coast— where utilities burn imported oil—you are experiencing the steepest price hikes. For instance, electric bills went up 42 percent in New York City in 1979 and the Midwest wasn't far behind. In fact, some people have found that even though they use both gas and electricity, the electric bill alone far exceeds their mortgage payments.

Nevertheless, consumer research tells us that "without having to sacrifice any major comforts or conveniences, you and your family can be able to save several hundred dollars each year on your energy bills, while helping to relieve the nation's energy crisis at the same time." How? One way is by being more energy-conscious when you buy and use your household equipment.

To begin with, let's get a clear picture of just where energy is used in our homes. Here, according to the experts, is how it breaks down: Space heating, 57 percent; water-heating, 15 percent; refrigeration, 6 percent; cooking, 6 percent; air-conditioning, 3 percent; clothes drying, 2 percent; other electrical, 11 percent.

It is very easy to see the three general areas in which savings can be made: heating and cooling; water-heating; and everyday use of appliances, primarily those concerned with the kitchen, such as refrigeration and cooking. In these areas, the amount of energy used is determined by two factors: the type of equipment or system you have, how efficient it is; and your normal pattern of energy use, whether economical or extravagant. You *can* change both factors to effect energy savings. Some energy expenses are a matter of personal preference. Others are a waste—which is quite foolish, and in these days of high prices and short supply, irresponsible. So, among the many comforts and conveniences derived from energy, now is the time to make wise choices.

How much can you actually save on your utility bills? With the suggestions given on the following pages of this book, it is estimated your savings could range from 20 to 40 percent of your annual energy bill. Thus, if your utility bills were about $1000, this could mean a savings of $200 to $400.

Some of the suggestions in this book involve investing money now for future savings. And, in the years to come there will certainly be more new energy-saving materials and appliances that may well cost extra. In considering such expenditures, we urge you to adopt the principle known as "life-cycle costing." Under this plan, the original cost of materials (or of appliances) is considered only one part of their total cost. Operating expenses during their estimated lifetime are the other.

In a sense, life-cycle costing is just another argument— provable in dollars and cents—for buying good quality. It is almost always less expensive in the long run. And, every rise we experience in the cost of fuel and electricity only underlines the value of life-cycle costing.

All prices quoted in this book were effective through 1980, and are subject to change.

The best way to conserve is to make energy thrift a part of our daily way of life—adopting the following practical, common sense energy habits. Conserving energy today not only means "waste not," want not but can also mean extra dollars in our pockets.

THE HOMEMAKER'S BOOK OF

Energy Savers

CHAPTER 1

Saving Energy in the Kitchen

THE KITCHEN is a good place to begin conserving energy. Though operating costs for major kitchen appliances are low when compared with some other home energy costs, a kilowatt saved in the kitchen can mean pennies and dollars trimmed from your utility bill.

Refrigerators

Food preservation is the third-largest consumer of energy within the family household budget (after space cooling/heating and hot water heating). Because the refrigerator operates day and night, it represents a large portion of the utility bill. Learn to coddle your refrigeration appliances. Granted, they use only a little energy each hour, but they do operate an average of 8,760 hours per year, and the costs add up quickly. At 5.5¢ per kilowatt hour, a seventeen-cubic-foot frost-free model uses $110 worth of electricity per year.

Buying a refrigerator Wise selection of a refrigeration appliance is a good place to start saving energy dollars. Look for a well-insulated cabinet and choose the **smallest** model refrigerator–freezer for your needs. A large model demands more energy than a small one. If it is too large, you won't be able to keep it filled to capacity, making the appliance work even harder.

1

Because less cold air is allowed to escape, a **two-door** re-frigerator–freezer uses less energy than the single-door type. When buying a new refrigerator, be sure to take **operating costs** into consideration. A frost-free model refrigerator will cost you approximately $40 more when you buy it, and about $21 more to run annually. Also, if you buy one of the new, high-efficiency models, you could, over a sixteen-year machine life expectancy, save as much as $600 in electricity bills. If yours is a manual-defrost model, don't let the frost build up on the freezer compartment to more than a quarter-inch thickness.

Refrigerator–freezers with water and ice dispensers use more energy for those features.

When buying that new refrigerator, it is energy economical to buy one with a **power-saver switch.** Most refrigerators have heating elements in their walls or doors to prevent "sweating" on the outside. In most climates, the heating element does not need to be working all the time. The power-saver switch turns off the heating element. By using it, you could save about 16 percent in refrigerator energy costs. Use it only when the air is extremely humid.

Locating your refrigerator   Place your refrigerator–freezer in a **cool location** for maximum efficiency. And avoid placing it against an uninsulated wall facing the sun, for such locations will make your appliance work harder and thereby consume more energy.

Provide adequate **air circulation** behind, above, and along the sides of the refrigerator and freezer.

Operating your refrigerator   The **seal** between the door and the body of the refrigerator can dry out and shrink in time. Put a dollar bill between the gasket that seals your refrigerator door and the refrigerator itself. Close the door with normal force; then pull the dollar bill straight out. There should be a slight drag. If not, your gasket is allowing cold air to leak out. This makes your unit work harder and use more energy. It is time to fix or replace the seal.

While most refrigerators and freezers operate more eco-nomically when **fully loaded,** don't overcrowd your appliance so as to interfere with circulation of air and thus allow development of "hot pockets." Keep the refrigerator well stocked but not overfilled. The motor in a half-empty refrigerator runs longer than in a full one. And it's harder on your motor to cool air space than to cool chilled foods.

Set the controls so your refrigerator maintains a temperature between **37 and 40 degrees,** and the freezer at **zero** degrees Fahrenheit.

Unless a recipe requires fast chilling or safety suggests it, the experts tell us to let **hot foods** cool slightly before putting them in the refrigerator or freezer.

Cover all **liquids** stored in your refrigerator. Moisture is drawn into the air from uncovered liquids, making the refrigerator work harder to keep its interior cold.

Clean refrigerator **condenser coils** at least every other month, using a vacuum with a nozzle attachment. Unplug the unit before cleaning the coils.

Unplug that old refrigerator if it is used only occasionally or to store marginal items. It may be costing you up to $100 a year if its gasket is worn or the compressor is bad.

Freezers

Buying a freezer   **Chest freezers** are more economical to operate than upright models because less cold air can escape from them when they are opened. However, the inconvenience of a chest-type model might outweigh the extra expense for you, especially if you are short, or would seldom get to the bottom of a chest freezer.

Freezer operation   It also pays to keep your freezer **well stocked.** Electric costs will be less, because each item in the freezer acts like a large, sub-zero ice cube.

However, when **freezing foods,** never place more than two or three pounds of food for each cubic foot of freezer size in any one day. Overstocking will raise the temperature and slow the freezing time.

Do you use a lot of **ice** in the summertime? Try emptying the ice trays into a plastic-foam cooler chest, and put the chest out on the counter. This way, the freezer door won't be opened every time someone wants ice, and it will keep making new cubes a lot faster.

Stoves

Stovetop cooking   Americans use more energy to prepare their food than farmers use to produce it. If you leave the **lids** off pots and pans while cooking, you are losing a lot of heat and speed in cooking—and you are wasting money.

If you have a gas stove, make sure the **pilot light** is burning efficiently, with a blue flame. A yellowish flame indicates an adjustment is needed.

Choose the pot or pan to **fit** the surface unit needed. A six-inch pan on an eight-inch unit wastes heat.

Cook fresh vegetables and fruits in tightly-covered containers in a minimum of water, and **reduce the heat** as soon as the steaming point is reached.

Use "high" heat only to bring foods to steaming, then switch to the **lowest heat** needed to complete cooking. Use "simmer" surface unit control for melting shortening, butter, or chocolate, and for reheating foods or keeping food warm.

After water boils, turn the control to "simmer" to maintain boiling temperature. Slightly boiling water is the same temperature as water that is boiling vigorously.

If you see flames licking around the sides of a pan on your stove, you are wasting gas. A gas burner should glow with a sharp, clear blue flame. If the flame has traces of yellow, it is turned up too high.

Approximately half of the gas that is used in a gas stove is used to fuel the pilot light, because pilot lights burn continuously. At least 10 percent of natural gas consumption goes to keeping pilot lights burning. For gas ovens and, incidentally, for all gas-operated appliances, switch-operated **electric starters** can be substituted for constantly-burning pilot lights. Or, look for an appliance with an automatic (electronic) ignition system instead of a pilot light. You will save up to 30 percent on gas.

Keep the **heat reflector** below the stove heaters clean, so you can get the maximum reflection of heat.

If you cook with electricity, get in the habit of **turning off the burners** several minutes before the allotted cooking time. The heating element will stay hot long enough to finish the cooking for you without using more electricity.

Oven operation   Did you know oven cooking is less expensive and conserves more energy than cooking on top of the range? The reason is that ovens are only on part of the time when they are used; their insulation allows them to coast part of the time. Burners stay heated the whole time they are on.

If your range has a **programmed oven,** you are lucky. They automatically turn the gas down to avoid overcooking, yet they keep the food at serving temperature.

Don't preheat the oven for foods that will cook at least an hour. (Exceptions: Preheating is always preferred for pastries, cakes and soufflés.) When you do preheat, ten minutes is enough. And, preheat your oven by setting the temperature at the heat you intend to use. A higher temperature gains nothing, because the oven won't preheat any faster.

Once you turn on the oven, make the most of it. **Cook the whole meal** in it, or include several dishes for future meals or desserts for the week. And, don't open that oven door unless you must. Each time you do, the temperature temporarily drops 25 degrees Fahrenheit.

If you move food directly from the freezer to the oven, especially roasts, you waste one-third of the energy you will use to get the roast properly cooked. **Let the food thaw** before you put it in the oven.

Use heat-treated **glass or ceramic pots** and pans in a conventional oven, to enable you to lower the heat by up to 25 degrees without sacrificing efficiency.

When roasting meat or baking casseroles, **turn the oven off** about fifteen minutes before the food is done. The retained heat will finish the cooking for you at no extra cost.

Those **nails** you buy to speed up baking potatoes work as

well on the grill as in the oven. They also work beautifully for baking acorn squash. Insert two of the nails into the squash before placing it in the oven, and it will really cut the baking time.

If you have a **self-cleaning oven,** this feature consumes an extra $4 worth of gas a year. Weigh this against the cost of oven cleaner bought during the course of the year, and you will probably find you are quite a bit ahead by purchasing the no-work, self-cleaning model.

When activating the self-cleaning cycle on an electric oven, start the cycle right after cooking, while the oven is still hot.

Frequent oven clean-ups help your oven to operate most economically. Do not line the oven with **foil;** this interferes with proper heating. **Continuous-clean ovens** do not consume extra energy and are a good alternative to consider if you plan to purchase a new oven.

The Peoples Gas Company of Chicago has issued a warning: "Use your gas range for cooking but not for **heating,** which is both expensive and dangerous. When a stove is used for cooking, it is operated only for a short time, then allowed to cool. When the oven is used for baking, the oven door remains closed. That conserves heat so the thermostat can reduce the gas flame for a time when the desired temperature is attained. When the oven door remains open, the oven must consume gas at full volume in its struggle to reach the temperature set on the thermostat— a costly process. The entire range, being metal, can become overheated, with possible damage to the appliance and to surrounding floors and walls, and injury to persons coming in contact with the hot surface."

What are the new **convection ovens?** In convection cooking, a blower forces heat around the food. You don't need to preheat the oven, which cooks food at lower temperatures and saves you both time and energy. (However, only a few gas range manufacturers today offer convection ovens.)

Microwave ovens   If you have a microwave oven, you're on the right track to saving energy. For instance, a half-inch hamburger patty can be cooked in three minutes in a microwave, costing only a fraction of a cent. The microwave oven is a stingy user of energy. It works on ordinary household current (110 to 120 volts), and **cooking times are very short.** The cost of operating a microwave oven is usually one-fourth to two-thirds less than that of the conventional electric range, which of course, must have its own 220-volt line.

Do not operate your microwave oven if any object is caught in the door, if the **door** does not close properly, or if the door latch, hinge, or sealing surface seem damaged, even slightly. People with pace-makers should *never* be in the same room with a microwave.

Keep the microwave oven and door scrupulously **clean.** Do not allow grease, crumbs or even bits of paper to remain stuck to the sealing surface of the oven door. If you suspect the door is not sealing properly, or that the oven isn't behaving as it should, have it checked immediately before further use.

Dishwashers

Your automatic dishwasher may actually **save water,** say the experts. A full cycle in your dishwasher consumes eleven to sixteen gallons of hot water, while hand-washing the dishes consumes nine to fourteen gallons. If you do hand wash, stopper the sink or use a dishpan. A running-water wash or rinse will use about thirty gallons of water per meal.

It is estimated that if every dishwasher user in the United States eliminated just one load a week, we would save the equivalent of about nine thousand barrels of **oil** each day—or enough to heat 140,000 homes during the winter months.

When buying a dishwasher, look for a model with air-power and/or overnight **dry settings.** These features automatically turn off the dishwasher after the rinse cycle. This can save you up to one-third of your total dishwashing energy costs, since turning off the dry cycle on your dishwasher saves 45 percent of the energy consumed by your dishwasher. And, during the winter months, if you **open the door** of your dishwasher as soon as the cycle is finished, it will not only add warmth to a chilly kitchen, but provide much-needed moisture to the air.

Try to eliminate running the dishwasher after every meal. Use it only when **filled to capacity**. Check with your city water department, regional agricultural agent, plumber, or other authority to find out how hard the water is in your area. This determines the amount of **detergent** you will need. A chart in the manufacturer's instruction booklet will tell you how much detergent to use with the type of water you have.

Remember, dishwashers can save energy if used intelligently. It may *sound* like a vast amount of water is surging through your dishwasher, but each two-gallon fill is actually being recirculated—at the rate of forty-five gallons per minute.

Other Kitchen Energy Savers

When rinsing dishes by hand or scrubbing vegetables, never let the **faucet** run continuously. This can waste five to six gallons a minute.

Install an **aerator** in your kitchen sink faucet. By reducing the amount of water in the flow, you will use less hot water and save the energy required to heat it.

Water Savers

If you install an **aerator** on kitchen and bathroom faucets, your warm water consumption will drop by two gallons per day. If you have an electric water heater, your electric bill will drop approximately $10 per year as a result.

Use cold water rather than hot to operate your **garbage disposal**. This saves the energy needed to heat the water, is recommended for the appliance, and aids in getting rid of grease. Grease solidifies in cold water and can be ground up and washed away.

Use your kitchen **ventilating fan** sparingly. In just one hour this fan can blow away a houseful of warmed or cooled air. Turn fans off just as soon as they have done their job.

If your kitchen lighting is incandescent, consider replacing it with **fluorescent lights.** One 40-watt fluorescent tube provides more light than three 60-watt incandescent bulbs. That alone can shave almost $1 per month off your electric bill. For more on fluorescent lights, see Chapter 13.

Pay attention to saving energy every day. Remember that the 100-watt bulb you leave burning overnight requires the energy equivalent of a pound of coal mined and burned. Equally important, consumers should pressure manufacturers to produce the most efficient appliances that will not only save money for the individual but conserve a national resource for everyone.

CHAPTER 2

Your Small Appliances

How CAN YOU save energy and money while using your small appliances? The best advice is to start by reading, or re-reading the **use-and-care manuals** that came with your appliances. They describe exactly how to use each appliance most efficiently. If you have lost any of these manuals, write to the company and ask for another (be sure to include the appliance model number). The use-and-care manual will also tell you if any regular maintenance is required, such as cleaning, adjusting, oiling.

A properly working product is clearly more efficient than one that is out of kilter. Proper care and maintenance of appliances ensures not only maximum efficiency, but also less energy use. **Clean** or replace air filters on appliances such as humidifiers or exhaust hoods. Clean and lubricate appliances and machinery when directed in your owner's manual. Vacuum cleaner dust bags should be emptied and replaced frequently, otherwise the full bag will reduce suction and increase vacuuming time.

Which Appliances Are More Efficient?

If you want to cut down on your utility bills, or at least try to keep them from escalating you into bankruptcy, it is essential to know **which appliances use the most energy.** For example, heat-producing appliances such as a toaster-oven or an electric frying pan use more electricity than small, motor-operated appliances such as your vacuum cleaner or blender.

However, before you start turning down and turning off, there are some small appliances that can actually be energy savers when they are used in place of a larger appliance. Electric skillets, toasters, waffle irons, popcorn poppers, electric fondue pots, and coffee pots generally use less electricity than a range doing the same jobs. Or, try baking a potato, chicken pot pie, or pizza in the toaster-oven instead of heating up your entire oven. It is not unusual for a full-sized oven to draw about 3,500 watts of power, while small electric toaster-ovens use only half that amount. This explains why small appliances, such as bean-pots, usually require less energy than comparable range cooking even though they require relatively high wattage.

When using these appliances, keep them out of **drafts,** which reduce their efficiency, and don't forget to turn them off when their job is finished. To avoid such things as the "on-all-day-coffee-warmer syndrome," store your extra hot coffee in a thermos bottle.

The **size** of the appliance is not directly proportional to the amount of energy it consumes. In fact, size has little to do with energy use. An electric wall clock may require 2½ watts and a hair blow-dryer 1200 watts. To assist consumers in gauging appliance energy use, the Commerce Department recently proposed that the appliance manufacturers voluntarily label their products to inform buyers of the wattage of electricity required for products. So, when selecting appliances, look to see if the energy demands and efficiency are indicated.

Some small appliances save in still other ways. For instance, an electric shaver takes a year to use up as much energy as it takes to heat one week's supply of hot water for lather shaving.

Televisions

Don't use energy-consuming special features on your appliances if you have an alternative. For example, don't use the **"instant-on" feature** on your television set. "Instant-on" sets, especially the tube types, use energy even when the screen is dark. And the "instant-on" device remains on twenty-four hours a day, which can add $25 a year to electricity costs. Use the "vacation switch" if you have one, to eliminate this waste, or have your TV serviceman install an additional on-off switch on the set itself or in the cord to the wall outlet.

Or, eliminate this waste by plugging the set into an outlet controlled by a **wall switch:** You can then turn the set off with the switch. Of course, turn off any TV set when you have finished viewing. If, however, a TV is turned off and on repeatedly, the heating and cooling wears out the tubes. So, if there is a short interval between shows you want to watch, leave the set on.

Look for "100-percent **solid state**" televisions, radios and stereos. They require about one-third less energy than conventional sets because they produce less heat, thereby putting a lighter burden on their cooling systems.

Irons

When you plug in your electric iron, think of ten 100-watt bulbs, as irons consume at least that amount of electric current. So, make your ironing short and don't start it up until you have more than a few pieces to do.

Remove all clothes from your dryer as soon as the cycle is finished, or even a little before, while they are slightly damp, to cut down on your ironing. Then, **do all your ironing at once.** Divide your clothes into groups depending on the amount of heat they require for ironing. Even the latest models of irons heat much more quickly than they cool. To save time, electricity and possible damage to fabrics from a too-hot iron, **start with fabrics requiring the lowest ironing temperatures** and work up to those needing the highest. Also, turn off your iron when you are called away, and turn it off slightly before you are through ironing so you can glide by on the retained heat.

Electric Blankets

Should you invest in an electric blanket? What about cost? Some electric blankets on the market today cost less than pure wool conventional blankets of the same size, and only about twice as much as a good synthetic blanket.

Safety   All but the cheapest double-bed electric blankets on the market today have **individual controls,** one for each side of the bed. If you fear sleeping under electrically heated wires, you needn't worry. Today's electric blankets are extremely **safe.** The electric wiring is encased in flexible watertight insulation that will not leak current, and the blanket's temperature cannot rise high enough to do any damage.

However, if a blanket is folded over several times, it can overheat, and manufacturers warn against this. They also build in about 9 safety thermostats, distributed along the 110 to 160 feet of heating wire, to control maximum temperature and keep it even throughout the blanket.

Using your blanket   The electric blanket will let you cut down on your heating bills by applying the crafty Japanese philosophy of **heating people, not rooms.** When using one, you can turn down the thermostat and save on today's expensive energy. When you consider that fuel oil prices are pushing $1.45 a gallon in some parts of the country, and that it can cost over $400 a month to heat a big house in the winter, the price difference between a conventional and an electric blanket becomes significant.

Typically, a **thermostatic control** on each blanket is marked with numbers from 9 to 14. These are rather meaningless numbers, they are not temperature settings. You are left to trial-and-error experimentation to discover which setting best suits your needs. Due to manufacturing inconsistencies, one blanket might keep you perfectly comfortable at the Number 4 setting, while a seemingly identical blanket would require a setting of 8 just to keep you warm. If the blanket you buy doesn't keep you warm enough, or if it keeps you too warm on the lowest setting, return it immediately for adjustment or repair under the warranty. You may have a lemon, and manufacturers are usually quick to fix or replace faulty units.

Electric blankets draw from 130 to 190 watts of **current,** depending on the size and brand. So, you can use one for the average cost of keeping a three-way 150-watt lamp burning all night. This isn't much, as you may have noticed if you have been trying to cut down on your electric bill merely by using less lighting.

The **timer** should glow brightly enough so you can see its settings in the dark, but not so brightly that the glow will keep you awake. It should also have an on/off indicator-light bright enough to be seen in a fully-lighted room, so that you won't accidentally leave the blanket on all day.

Wash your electric blanket only according to the manufacturer's instructions, which usually call for lukewarm water and mild soap. Do not dry clean. Cleaning chemicals can damage the important heating wire insulation. And, do not store these blankets in **moth balls**. Electric blankets are mothproof. And mothball fumes may damage electrical insulation.

In Yard, Garden and Workshop

Buy **power tools** that use the lowest power sufficient for the work you need done.

Keep the **cutting edges** of all your electrical tools as sharp as possible, so they operate at maximum speed and use relatively less power. Properly lubricate and clean all your electrical tools so they require minimum power.

Use **hand tools,** hand lawn mowers, hand clippers, pruners, etc., when they are adequate for your purposes.

Before leaving the workshop or bench, make sure the soldering iron is off, the shop lights are out, and all heating devices are off.

Never allow the machinery in your gas-powered equipment to **idle** for any length of time. Get into the habit of turning off the engine when finished with one job, and turning it back on when ready for the next.

How does your garden grow? Did you realize that even your little vegetables and bushes could be consuming expensive energy? They are, if you use **artificial fertilizer.** Petroleum and natural gas are generally used as raw materials and for fuel in the manufacture of artificial fertilizer. So, save money and energy and make your own fertilizer by composting leaves, lawn clippings, organic garbage, and other organic waste matter.

Energy Chart

The chart at the end of this chapter tells you the **average operating costs per year of some seventy appliances,** both large and small, in the average home. You can also use this chart to figure out how much it costs to run each of these appliances per year in your home. Just multiply the average number of kilowatt-hours (kwh) used annually, as listed in the first column of the chart, by your local electric rate. (The chart uses an average of 5 cents per kilowatt-hour, but rates throughout the country can go anywhere from 4 cents to 6 cents and more). Sometimes the rate is shown on your electric bill. If it is not, divide the total amount by kilowatt-hours, or call your utility company for the rate.

Using some appliances can actually **save you money.** A two-slice toaster, for instance, has a high wattage of 1100, but toasting bread in an oven would require three times as much energy. Average annual toaster usage is 40 kwh per year. If your electric rate is 5 cents per kwh, multiply $.05 × 40 = $2.00, which is what it costs you to toast bread each year.

The figures on the chart are **national averages.** For instance, we show the Hairsetter/curler being used 14 kwh per year at a cost of $.70. However, if you have four or five girls using the appliance at your house, your cost would be multiplied by this figure.

Here is still another method of figuring the kwh if you think you use your appliances considerably more or less, or that their wattages are higher or lower, than what we have shown here. Find the wattage of the appliance by checking the name plate. Then multiply the watts by the number of hours you estimate you use the appliance, and divide by 1,000. For example, if you use a 75-watt light bulb for 10 hours, the kwh work out to $75 × 10/1000 = .75$ kwh. If the electric rate in your area is 5 cents per kwh, it would cost you 3.75 *cents* to operate the 75-watt bulb for 10 hours.

Energy Requirements of 70
Home Appliances[a]

	Est. KWH Used Annually	Cost per Yr. at 5¢ KWH
Air cleaner	216	$ 10.80
Air Conditioner (room)	2,000	100.00
Attic Fan	291	14.55
Blenders and Food Processors	270	13.50
Broiler	100	5.00
Can Opener	1	.05
Carving Knife	8	.40
Chafing Dish	9	.45
Circulating Fan	43	2.15
Clothes Dryer	1,200	60.00
Coffee Maker (brew cycle)	106	5.30
Coffee Maker (warm cycle)	48	2.40
Corn Popper	9	.45
Deep Fryer	83	4.15
Dehumidifier	377	18.85
Dishwasher (includes energy used to heat water)	2,100	105.00
Dishwasher only	363	18.15
Egg Cooker	14	.70
Electric Blanket	197	9.85
Floor Polisher	15	.75
Fluorescent Light (three fixtures)	260	13.00
Fondue Pot	9	.45
Food Processors	270	$ 13.50
Freezer (15 cubic feet)	1,195	59.75
Freezer (15 cubic feet, frostless)	1,761	88.05

NOTE: When using these figures for projections of energy use for your appliances, such factors as the size of the specific appliance, the geographical area of use, and individual use should be taken into consideration.

[a] Table compiled with the help of:
   Electric Energy Association
   Citizens Advisory Committee on Environmental Quality, Washington, D.C.

Energy Requirements of Appliances, cont.

	Est. KWH Used Annually	Cost per Yr. at 5¢ KWH
Frying Pan	200	10.00
Germicidal Lamp	141	7.05
Griddle	50	2.50
Hair Dryer (per person)	26	1.30
Hair Setter/Curler	14	.70
Heat Lamp (infrared)	13	.65
Heater (portable)	176	8.80
Heating Pad	10	.50
Hot Plate	90	4.50
Humidifier	163	8.15
Iron	144	7.20
Light Bulbs (per eight 75-watt bulbs)	865	43.25
Mixer	13	.65
Microwave Oven	190	9.50
Radio	86	4.30
Radio/Record Player	109	5.45
Range with Oven	1,550	77.50
Range with Self-Cleaning Oven	1,205	60.25
Refrigerator (12 cubic feet)	728	$ 36.40
Refrigerator (12 cubic feet, frostless)	1,217	60.85
Refrigerator–Freezer (14 cubic feet)	1,137	56.85
Refrigerator–Freezer (14 cubic feet, frostless)	1,829	91.45
Roaster	205	10.25
Rollaway Fan	138	6.90
Sandwich Grill	33	1.65
Sewing Machine	11	.55
Shaver	1.8	.09
Sun Lamp	16	.80

Energy Requirements of Appliances, cont.

	Est. KWH Used Annually	Cost per Yr. at 5¢ KWH
Television:		
Black & White, Tube Type	350	17.50
Black & White, Solid State	120	6.00
Color, Tube Type	660	33.00
Color, Solid State	440	22.00
Toaster	50	2.50
Toothbrush	0.5	.025
Trash Compactor	50	2.50
Vacuum Cleaner	46	2.30
Vibrator	2	.10
Waffle Iron	22	1.10
Washing Machine — Automatic (inludes energy used to heat water)	2,500	$125.00
Washing Machine only	103	5.15
Washing machine (non-automatic, includes energy to heat water)	2,497	124.85
Washing machine (non-automatic)	76	3.80
Waste Disposer	30	1.50
Water Heater	4,219	210.95
Water Heater (quick recovery)	4,811	240.55
Window Fan	170	8.50

CHAPTER 3

Water-Heating and Saving Tips for Laundry Room and Bath

A PRINCETON UNIVERSITY study recently showed that "there can be a 50-percent difference in utility bills between families living in identical town houses." Water-heating plays a major role in these differing expenses. Next to heating and cooling systems, water-heating is the largest user of energy, representing about **15 percent of the family's energy budget.** So, it makes sense that careful use of water and proper maintenance of heaters will mean considerable energy and money savings.

Water-Heaters

Perhaps most important is the fact that the water-heater is the one appliance that is generally installed, set, and forgotten. However, there are quite a few ways you can relieve the workload and economize with your water-heater.

• **Select the proper size heater** for your family needs. If the heater is too large, energy is spent heating unneeded water. Naturally, if it is too small, you will not get all the hot water you need daily. The consumption of hot water is affected by the number of family members, their ages, habits, and home facilities. Studies show the usual minimum hot water needs for a family of four persons with one bathroom, an automatic clothes washer, and an automatic dishwasher, would be approximately thirty gallons per day.

• When possible, **locate** your water-heater as close to the point of hot water use as possible. Long pipes retain water, which cools off. This means using cold water and waiting for hot water. So, you lose heat and waste water.

• For supply pipes, use the **smallest diameter** possible in order to reduce heat-loss and the volume of trapped water. When you can't avoid long pipe lines, insulate them to minimize heat dissipation. In sprawling homes, it may be more economical to have two water-heaters located in the main water-using areas instead of one heater centrally located.

• If you **go away** for an extended period of time, turn down or turn off your water-heater. If you are turning off the gas pilot light, don't forget to re-ignite it before turning the heater back on.

• Do you know how high to set your water-heater and still economize on energy? If you have a dishwasher, set the heater at 140 degrees for best results. More is wasteful. If *you* are the dishwasher, set it at 120 degrees. And, the U.S. Department of Energy suggests we **turn down that water-heater thermostat,** as they estimate you can save at least $20 a year by lowering the setting on your electric water-heater from 150 to 130 degrees. (Savings will be less for natural gas heaters.) The higher the temperature inside a water-heater, the more heat that's lost through the walls of the tank.

If you are uncertain about the tank water temperature, draw some water from the heater through the faucet near the bottom and test it with a thermometer.

• **Insulate** your water-heater. By wrapping a piece of thick, aluminum-backed fiberglass insulation around your tank, you can reduce the amount of heat lost through its sides and save as much as $20 a year. However, be sure not to block off needed air vents, especially with oil or gas water-heaters. The insulation materials should not cost more than $10 if you do the work yourself.

• **Drain** your water-heater tank. Studies have shown that sediment collecting at the bottom of a water-heater reduces the system's efficiency. Every few months, open the plug at the bottom of the tank and let the water flow out until it runs clear.

Average Hot Water Used per Day

2 adults and 1 child	60 gallons
2 adults and 2 children	70 gallons
2 adults and 3 children	80 gallons

• **Insulate the pipes.** In many homes there can be sixty feet or more of three-quarter-inch piping between the hot water tank and the faucet. That length of pipe contains up to two gallons of water. You must run off that two gallons, plus an additional gallon, to warm up the pipes, before any really hot water reaches its destination. If a hot water faucet is used ten times a day, this means you may be wasting nine hundred gallons of hot water a month. The cost of insulating sixty feet of pipe would pay for itself in six months!

• When buying a new water-heater, remember that energy-efficient water-heaters may cost a little more initially, but reduced operating costs over a period of time can more than make up for the larger outlay. Select one with **thick insulation** on the shell.

Clothes Washers and Dryers

Washing clothes Energy studies show we should wash clothes in **warm or cold water,** using hot water only when essential. If everyone washed clothes in warm or cold water, we'd save the equivalent of 100,000 barrels of oil a day, enough to heat 1.6 million homes through the winter.

And, based on the Washington, D.C. fuel rate, studies show you will save about 4 cents a load if you wash on the cold-water setting instead of the hot. (This might be fine if clothes have light to moderate soil, but perhaps it will be worth the extra 4 cents to get heavily soiled clothes really clean.)

The laundry experts also tell us to save energy by **rinsing clothes in cold water.** A cold-water rinse is as effective as warm, since rinsing is basically a dilution process. And, did you know that a cold-water rinse uses approximately a third of the energy of a warm-water rinse? In fact, a cold-water rinse is actually recommended for most permanent-press fabrics. Hot water, as you well know, shrinks cotton materials. By using warm-water wash and a cold-water rinse, you will save 60 percent in energy consumption while cutting shrinkage.

Fill your clothes washer and dryer (unless there are special instructions to the contrary), but do not overload. If every household cut the use of its clothes washer–dryer by 25 percent, the nation's savings would be equivalent to 35,000 barrels of oil a day, enough to heat more than 400 *billion* gallons of water a day, say the experts.

And, since small loads require much less water than big ones, always use the **water level control** on your washer.

For heavily soiled garments, use a pre-soak product or the soak cycle on your washing machines. **Reduce washing time** and protect your fabrics by pre-treating stains when they are fresh. You may have to wash these clothes only once instead of twice.

Be sure to follow detergent instructions carefully in your laundry (and dishwasher, too). **Oversudsing** makes your equipment work harder and consume more energy than necessary.

If you have a **water-softening system,** check to see how often your equipment regenerates and backwashes. It can use as much as 100 gallons of water each time it does this. You may want to cut down on the use of such equipment. Reserve softened water for kitchen use, bathing and laundry. Use unsoftened water for all other purposes. This may require a bypass line for your water-softener, but this is advisable under any circumstances.

Buying a washer or dryer   Whether you are buying your first washer and dryer or replacing old ones, here are some guidelines from the experts to help you choose wisely, while saving energy and money in the long run:

1. **Read the product literature** thoroughly to determine how much water is needed for a complete cycle, comparing brands and models. You will find the amount will vary greatly,

and over a period of time you can stack up substantial savings on hot water by choosing a model that uses water sparingly. Why do some machines use less water than others? Because the water-misers have less space between the perforated inner tub and the outer tub.

You will find automatic washers also have motors varying from one-third to one-half horsepower, but they consume very little energy. The key to savings is to buy a washer with **flexible controls** that can be used to conserve energy in many ways.

2. Water-level controls save water on small loads. And, separate wash- and rinse-water **temperature controls** allow cold-water rinses on all cycles.

3. An **automatic soak cycle** will enable you to use cooler water for washing, and also helps with low-phosphate detergents.

4. A **bleach dispenser** makes it easy to boost your cleaning power, especially when you are using cooler water.

5. A machine with a **super-fast spin cycle** will extract more water and reduce drying time. Such cycles are helpful on terry towels, blue jeans, etc., but should not be used on permanent-press fabrics.

6. Washers with a **suds saver** allow you to set controls to automatically divert wash water into a separate storage tub where it is stored during the rinse-and-spin portion of the cycle. When you want to reuse wash water, adjust the controls, and the storage tub automatically pumps the water back into the washer during the next cycle. You can save the most energy by reusing the hot water from the tub, since heating of water represents 95 percent of all the energy used to wash clothes.

7. Energy-efficient **electric ignition pilot lights** are beginning to replace constant-burning pilot lights on gas dryers. You will also have to decide whether to buy an inexpensive timed dryer or the more expensive **sensor** type. The sensor tests the amount of moisture in the clothes and automatically terminates the cycle when they're dry, while the electronic control is more accurate and more expensive than the automatic dry control on older models. Guessing the time needed to dry clothes in a timed dryer can result in overdrying, which not only wastes energy but wears out the clothes.

Your dryer **Consolidate your loads** so that you dry a full load at a time. In other words, fill the clothes dryer but do not overload it.

Keep the **lint screen** in your dryer clean. Remove lint after *each* load. Lint impedes the flow of air in the dryer and requires the machine to use more energy.

Exhaust your dryer to the outside. A dryer that exhausts moisture-laden air indoors tends to recirculate that air through the dryer, lengthening drying time and requiring more energy. Also, keep that outside exhaust of your clothes dryer clean. Check it regularly. A clogged exhaust lengthens the drying time and increases the amount of energy used.

If your dryer has an **automatic dry cycle,** use it. However, overdrying merely wastes energy. Dial the degree of dryness you want. Clothes don't always need to be crinkly dry. Some dampness is desirable for easier folding and for fewer wrinkles in many fabrics.

Dry your clothes in **consecutive loads.** Stop-and-start drying uses more energy because a lot goes into bringing the dryer up to the desired temperature each time you begin.

And, **separate** the drying loads into heavy and lightweight items. Since the lighter ones take less drying time, the dryer doesn't have to be on as long for these loads. If drying the family laundry takes more than one load, leave small, lightweight items until last. You may be able to dry them, after you turn off the power, with heat retained by the machine from earlier loads.

Ironing

Plan your ironing sequence **from low to hot,** as suggested in Chapter 2. Even the latest-model irons heat much more quickly than they cool, so start with fabrics that require low heat. You not only save time and electricity, but there also will be less possible danger of damaging fabrics from a too-hot iron. Work up to the fabrics that need the hottest or highest setting.

Remove those clothes that will need ironing from the dryer while they are **still damp.** There is little point in wasting energy by drying them thoroughly if they only have to be dampened again.

You can also save ironing time and energy by "pressing" sheets and pillow cases on the warm **top of your dryer.** Fold them very carefully, then smooth them out on the flat surface.

How about saving energy needed for ironing by hanging certain items of clothing in the bathroom while you are bathing or showering? The **steam** often removes the wrinkles for you.

In the Bathroom

The experts say that 70 percent of the water inside the average home is used in the bathroom, where one of history's biggest water consumers is perched—the toilet. Although only 2½ gallons are actually needed, the typical American toilet uses anywhere from five to eight gallons per flush—and some of us flush as though it were the only way to discard things. Throw in a facial tissue—flush; throw in a cigarette butt—flush. Each day we flush away 29 gallons of water per person, more than five times the amount a person living in an underdeveloped country uses for all purposes in a whole week!

One way to conserve is by filling a couple of plastic bottles with sand and put them in the toilet tank to displace water. This will reduce the amount of water that goes down the drain with every flush. Or, for a few dollars you can buy commercially made toilet tank dams from a plumbing supply outlet.

Check for silent leaks by putting food coloring in the tank and watching to see if it shows up in the bowl. Leaks can cause up to 95 percent of the complaints about excessive water bills.

You can also cut down on waste by not letting the water run while brushing your teeth. A family of four who lets the water run while brushing their teeth can waste eight gallons a day. And, if half the people in this country had that same habit, we would waste 160 billion gallons of water per year, just brushing our teeth.

Does taking a **shower rather than a bath** really save water? The U.S. Department of Agriculture offers some facts: "A bath normally requires 30–50 gallons of water, while a shower uses 5–15 gallons per minute. By setting your shower on low or on spray rather than full force, it is possible to take up to a 6-minute shower and still save water. And, a quick shower will usually use less than half as much hot water as a bath in a regular-size tub." (You can test this yourself by plugging the tub when you take a shower to see how much water you have actually used.)

Install a **flow restrictor** in the pipe to the showerhead. This easy-to-install device can save a considerable amount of hot water in a year. It costs less than $10, fits into the pipe to the showerhead, and restricts the flow to an adequate four gallons of water per minute.

The men in the family should not waste hot water by **keeping the faucet running** as they shave. They can pour as much as six gallons of hot water down the drain during a five-minute shave. Ask them to close the sink drain while they shave.

Leaky faucets   Replace all **leaky faucet washers.** If your hot water faucet leaks just one drop per second, it wastes 2,500 gallons a year. That's water you paid to heat. If you heat electrically, a 10-cent washer could save you more than $25 per year in electricity.

The experts further advise that a pinhole leak can waste up to 170 gallons a day; a fast drip, 970 gallons a day; a toilet leak, up to 3,000 gallons a day!

To give you an idea of the cost involved, the following table was drawn up with the cost based on 3.5¢ per kilowatt-hour—and in many areas of the country electricity is considerably more expensive:

Energy Costs of Leaky Water Faucets

Drops per Minute	Gallons per Month	KWH per Month	Cost Per Month	Cost per Year
60	192	48	1.68	$20.16
90	310	78	2.73	32.76
120	429	107	3.75	45.00

Toilet leaks   Would you believe that nearly half of all the water used in your household goes down the **toilet drain** at the rate of five to eight gallons per flush? Since it is doubly smart to be stingy with water, consider replacing your conventional toilet tank with a new flushing device that uses a mere two gallons of water to flush efficiently. The unit costs about $70, comes in approximately fifty colors, and can be installed by any fairly handy homeowner.

The U.S. Department of Agriculture cautions people on the practice of placing bricks in their toilet tank to reduce its volume and save water. Bricks tend to crumble when in the water for a long period, and could ruin the valves. The U.S.D.A. therefore suggests using a **filled quart plastic bottle** in the tank. You will save one quart of water per flush.

Avoid using the toilet as a trash basket for facial tissues, etc. Each flush uses five to seven gallons of water.

When buying a new toilet, look for a "low volume" model. They don't use as much water per flush.

How to Fix a Toilet

If water continues to run into the toilet bowl after flushing and the tank doesn't fill with water, check the stopper ball as follows:

1. Look into the tank at the stopper ball, which closes the water over-flow at the bottom of the tank, and joggle the flush handle. Is the stopper ball positioned directly over the tank-ball seat? If not, the **lift rod** may be bent. Straighten it and now see whether the stopper ball falls properly. If the lift rod is bent beyond repair, replace it with a new one. They are usually neatly packed and inexpensive.

2. The **stopper ball itself** may be worn and need replacing. Turn off the water, using the valve under or behind the bowl, or tie up the float arm so that the water flow stops. A simple way to do this is to place a stick of wood or a broom handle across the top of the open tank and tie the float arm to the stick.

3. Unscrew the stopper ball from the lift rod. Take it to the hardware store and get a new one the same size. It should cost approximately $2 or less.

4. Take some fine steel wool and rub around the tank-ball seat to scour it clean. Then screw on the new stopper ball and turn on the water. The toilet should now be working properly once again.

Now, the float ball is something else. If the tank is filling properly but water continues to run into the toilet bowl after flushing, the float mechanism may be at fault.

1. Lift the **float arm** to see whether the water stops leaking. If it does, either you need a new float ball or the float arm needs adjusting. Note that the water level in the tank may not be correct. It should be within one inch of the overflow pipe. If the water level is lower than this, bend the float arm upward slightly. If the water level is high and sloshes over the overflow pipe, bend the float arm downward slightly.

2. If adjusting the arm fails, you may need to replace the **float ball.** Either turn off the water or tie up the float arm (see Step 2 above).

3. Unscrew the float ball from the float arm, take it to the hardware store, and replace it with a new float ball. If this solves the problem, it means that the float ball was leaking and no longer floated.

How to Fix a Leaky Faucet

It drips like a head cold and can be just as annoying. The cause is nearly always a worn out washer that costs about 2 cents, though a plumber may charge you $20 and up if he has to replace it for you.

You'd be wise to have a box of assorted-size washers on hand for just such an emergency. Before you do anything, however, turn off the water. Look for the **shut-off valve,** normally right under the sink. If it isn't there, find the main shut-off valve where the water pipe enters the house.

Water off, simply unscrew the handle. Undo the collar nut with a small monkey wrench, then another nut below it, and pull out the faucet stem. *Important:* **Lay out each part in order of removal;** reassemble in reverse order.

The culprit washer, looking cut and worn, will be at the bottom of the faucet stem. Remove with a screwdriver, replace with a washer the same size, then reassemble.

If you hear a banging or **hammering in pipes** whenever you turn off a faucet, don't ignore it and hope it will go away. No problem involving home maintenance will ever go away until it has been fixed. The banging is caused by excess vibration, and has been known to split pipes wide open. However, before you call the plumber, check the exposed lengths of pipe and their metal straps and mounting brackets. If a pipe is loose or sags, firmer support may arrest both the vibration and the noise.

CHAPTER **4**

Keeping Cool as a Cucumber
—for Less

TODAY, APPROXIMATELY one-third of the population lives in air-conditioned comfort. Thirty-five million Americans enjoy centrally-air-cooled homes, while another thirty-five million use room air-conditioners. Fuel shortages—together with inevitable increases in the cost of power—make cheap, efficient ways to cool your home attractive, both in conserving energy and in easing the family budget.

Purchasing Air-Conditioners

When purchasing room units you will find there are hundreds of models to choose from. How can you tell which one best meets your needs? With the knowledge of a few basic facts, and the help of an experienced salesman, picking out the right air-conditioner for you can be a relatively simple task.

Room air-conditioners are simple machines designed to cool, dehumidify and filter the air. They have two key parts: a cooling element, and a fan. The cooling element is much like the one in your refrigerator and consists of a compressor (pump) and a series of tubes through which the compressor pumps a cool refrigerant. The fan circulates the air, and a third part, the filter, removes dust and other particles from the air passing through it.

These three elements work together. The fan pulls the air from the room into the unit. The air moves through the filter and is cleaned before being swept across the cold compressor pipes, which cool and dehumidify it. Finally, the fan pushes the cool, dry air back into the room.

Estimating capacity   The most important feature of any air-conditioner is its capacity; it is necessary to purchase a unit that is **large enough** to adequately cool the room, but it is equally important to avoid buying an air-conditioner that has too high a capacity. If a room gets too cold, moisture will condense and the coils in the air-conditioner may freeze. The unit will stop producing cool air and will not dehumidify properly.

On the other hand, a unit that is **too small** must operate almost continuously to bring temperatures down, decreasing the life of the equipment while failing to cool efficiently.

Here is a chart to show you how to estimate the needed capacity of the unit you are about to buy. To use the chart, follow these directions:

1. Determine the floor area, in square feet, of the room to be cooled (multiply room width by length).

2. Depending on what kind of space is above the room to be cooled (see explanation below) select the applicable column (A, B or C) and the "exposure" sub-column, and find the floor area that most closely resembles yours. Then read across to determine the adjusted cooling capacity.

Character of Space above Room

Column A: The room above the room to be cooled is occupied.

Column B: An unoccupied attic and insulated ceiling are above room to be cooled.

Column C: Uninsulated ceiling and unoccupied room or attic is above room to be cooled.

3. If the room is to be cooled primarily at night, or after work, multiply the unadjusted cooling capacity by 0.7.

4. If more than two people usually occupy the room, add 600 BTUs per hour to answer in Item 3 for each person over two.

How to Estimate Capacity for Air-Conditioner

Column A		Column B		Column C		
Sunny Exposure (sq. ft.)	Northern Exposure (sq. ft.)	Sunny Exposure (sq. ft.)	Northern Exposure (sq. ft.)	Sunny Exposure (sq. ft.)	Northern Exposure (sq. ft.)	Unadjusted Cooling Capacity
100	130	80	100	60	70	4,000
150	200	140	160	90	110	5,000
220	280	200	230	130	170	6,000
300	370	260	290	180	210	7,000
380	450	320	380	220	260	8,000
460	540	400	450	270	300	9,000
540	630	460	520	310	350	10,000

5. Add 4,000 BTUs per hour to Item 4 if the area to be cooled includes a kitchen or indoor barbecue.

The result is your estimate of the needed cooling capacity.

Cost of operation   Economy is, of course, another vital consideration in selecting an air-conditioner. Air-conditioners range in price from approximately $150 to more than $500, depending on capacity and other factors. However, the least expensive unit may not be the most economical in the long run, and it could be better to purchase a more costly unit that gives you more cooling for your electricity dollar. The information necessary to compute the **amount of cooling per amount of electricity** is given on the nameplate of all air-conditioners. The capacity of an air-conditioner is measured in BTUs (British Thermal Units) per hour. (A BTU is a measurement of heat; one BTU is about equal to the heat given off by a wooden kitchen match burning to ash.) An air-conditioner that is rated at 8,000 BTUs per hour will extract 8,000 BTUs of heat from the air each hour.

The amount of electricity a unit uses during an hour of operation is also noted on the nameplate. This measurement is in **watts per hour.** To find out how efficient an air-conditioner is, simply divide the wattage figure into the BTU figure. The answer tells you how many BTUs of cooling you get for each watt of electricity used. For instance, if there are two 8,000-BTU–air-conditioners, and one uses 900 watts per hour while the other uses 1,200 watts per hour, the first will give you nearly 9 BTUs of cooling per watt and is more economical than the second, which gives you less than 7 BTUs per watt. Even if the first unit is more expensive to purchase, it could be cheaper in the long run because it will save you money on electricity.

Be sure to compare the energy-efficiency of units you are considering—usually listed as EER (**Energy Efficiency Ratio**). The EER rating (unlike BTUs, which are expressed in thousands), is a single number. The higher the number, the more efficient the appliance, and the easier it will be on your electricity bill.

Other things to look for   Before you purchase an air-conditioner, determine whether it is compatible with the **electrical wiring system** in your home. Most room air-conditioners use standard 115-volt household current, but some larger units require 208- or 230-volt current. Some localities have special regulations regarding air-conditioners, and certain safety measures to prevent overloading a circuit should be taken in any event. Therefore, it is best to have an electrician check the wiring in your home before you purchase an air-conditioner.

Also, check the **size of the unit**. Will the air-conditioner fit in your window? If it is large, will it block out too much light? (Light, portable units that can be carried from room to room are now available.)

Quietness of operation is especially important if the air-conditioner is going to be used in a bedroom. All room air-conditioners make noise, but some are quieter than others. You can compare noise levels at the dealer's before you buy. Usually, a smaller unit with the same BTU rating as a larger unit will make more noise because it contains less sound insulation.

Filters on most models are similar, in that they can either be cleaned with a vacuum cleaner or washed. However, check for easy accessibility and removal.

Almost all air-conditioners contain **thermostats** that, once set, keep the room at a constant temperature. If you purchase an air-conditioner without a thermostat, you must regulate room temperature yourself by turning the unit on and off.

An **adjustable fan** is available on many units today. Some have two fan settings and others have three. They give the unit more flexibility. If you want to cool a room quickly, you can turn the fan to high and get maximum output from the air-conditioner.

An **exhaust–fresh-air switch** is also available on many models. When the unit is set on "normal" it recirculates the air within the room. When the unit is set on "exhaust" it draws some air from the room to the outside. When the "fresh air" setting is used, the machine introduces fresh air into the room.

Grille design in an air-conditioner means more than adding beauty to the unit. If the louvers are stationary, they should point at an upward angle. If the louvers are movable, you can set them any way you wish.

Read the **warranties** thoroughly. The standard warranty covers the compressor for five years and other parts for one year, but there are variations.

The cost of keeping cool will be higher than ever this summer, you must use your air-conditioner wisely. A good investment for a working family is a special **timer** that will turn your units on in late afternoon.

Thermostat Settings

A study showed that an air-conditioner's thermostat setting has a great influence on **cost of operation.** In two homes of the same size and construction, for example, one thermostat was set to maintain 73 degrees while the other maintained 78 degrees. It was found that operating cost increased approximately 8 percent per degree of lower-maintained inside temperature. In some units, there is a control which enables the user to bring in a small quantity of fresh air from the outside. Under certain conditions, this may be desirable, but the heat and moisture must be removed from the new air added, thus increasing the cost of operation. Each degree the thermostat is raised cuts **5 percent** off your air-conditioning costs.

And, the experts contend it is both healthy and wise to make your home no more than **15 degrees** cooler inside than out, even if the outdoor temperature soars. On very hot days, set the air-conditioner fan speed at high. If it is very humid, however, set the fan at low speed to provide more moisture removal. Otherwise, select a medium setting.

If no one is home, be sure to turn the air-conditioning units **off.** If you can't bear to walk into a hot house, don't leave the air-conditioner on all day, but rather fix an automatic thermostat timer to the system so that you can pre-set it to automatically turn on one-half hour before you expect to arrive home.

Air-Conditioner Maintenance

Routine maintenance is simple and inexpensive, and you will find it will reduce operating costs and keep your units in peak working order. Use 10- or 20-weight **oil** to lubricate your fan motor. Also, check your fan belt for wear periodically.

Clean or check the **filters** once a month. You might replace your regular air cleaner with an electric one. Electric air cleaners use about as much electricity as a light bulb.

Keep your outdoor unit clear and **unobstructed by trees,** bushes and weeds. Fallen leaves, brush and grass can block the coils. Thoroughly clean the outdoor unit's grille, since clogged coils can cut efficiency. Under most circumstances, the spray from a garden hose can do the job. If chemicals are needed for cleaning, call a service man.

Miscellaneous Energy-Saving Ideas for Keeping Cool

Unshaded windows and other glass openings will greatly increase air-conditioning requirements. **Awnings** installed over windows reduce the sun heat penetrating into the room as much as 50 to 75 percent. If awnings are not practical, insulated draperies or venetian blinds will give good results, as well as louvered sun screens and shades.

Shrubbery against the outside walls of your home will also effectively reduce the amount of cooling required. However, as we have suggested, shrubbery planted too close to the air-conditioning unit itself will reduce the necessary air-flow and cause inefficient operation. **White roof tile** and pastel masonry wall paints do much to reflect the rays of the sun away from the interior of the house.

Be certain that your home is **well insulated** and sealed to keep cooled air inside. In addition to insulation, storm doors and windows or double-glazed windows will reduce the need for air-conditioning.

Make sure the **ducts** in your air-conditioning system are

properly insulated, especially those that pass through the attic or other uncooled places. This could save you almost 9 percent in cooling costs.

Don't set your **thermostat** at a colder setting than normal when you turn your air-conditioner on. It will not cool faster. But, it will cool to a lower temperature than you need and use more energy. Remember, if everyone raised his air-conditioner temperature just 6 degrees, we'd save the equivalent of 36 million kwh of electricity in the nation in one year!

Turn off the window air-conditioners when you leave a room for several hours. You will use less energy cooling the room down later than if you had left the unit running.

Consider using a **fan** with your window air-conditioner to spread the cooled air farther without greatly increasing your power use. But be sure the air-conditioner is strong enough to help cool the additional space.

Don't place **lamps or TV sets** near your air-conditioning thermostat. Heat from these appliances is sensed by the thermostat and could cause the air-conditioner to run longer than necessary.

If your home has window air-conditioners and you also have central heating, you may be sending some of your cold air out through the **fresh-air return vents** for the heat. Usually located at floor level, these vents can have a nice draft going through them. Even if they aren't the kind that can be closed, it is an easy matter to remove the grille and block these vents off until next winter.

Place window units on the **north or shady side** of the house to protect the unit from direct sunlight. Otherwise, the sun will increase the unit's work load. An awning can be used to shade the unit itself (just make sure the awning does not trap heat). Outdoor units should also be placed on the north side of the house.

Further, if your home is air-conditioned, you can save 15 percent on your air-conditioning costs by leaving the **storm panes** closed when the unit is operating.

Save your moisture- and **heat-producing activities**, such as laundry, bathing, and cooking, for cooler times of the day.

And, be sure to turn your air-conditioning off in **unoc-**

cupied rooms. It also helps to use fans in the attic and over the stove to blow out warm air that might compete with the air-conditioning.

Ventilating Fans

During the summer, the heat that builds up in the attic can rise to 135 degrees or more. This heated air can penetrate through the ceilings (even if they are insulated), warming the rooms below and putting a big load on the air-conditioner. Therefore, ventilating fans, properly selected and installed, can contribute to your comfort and cut cooling costs by reducing humidity and controlling heat buildup, according to many experts. Powered **attic-space ventilators** draw in outside air and flush it through the attic. This can greatly increase comfort in uncooled houses, they say. It also may significantly reduce the cost of cooling a home with air-conditioning. In fact, with proper attic ventilation you may be able to cool your home with a smaller-capacity air-conditioner.

Powered ventilators come in a variety of types and sizes to fit different roof and attic shapes. They are usually mounted on the roof, but can be installed on the wall of the attic if necessary. They run on regular household current, and cost about as much to operate as a 150-watt light bulb.

Ventilating fans may also cut cooling costs in your **kitchen, bathrooms and laundry area.** The fans not only add to your comfort by removing humid air, they also make air-conditioners operate more efficiently. Maintaining the proper humidity lessens the overall burden on your cooling system, and allows you to be more comfortable with the thermostat set a few degrees higher.

Whether or not you have air-conditioning, you should make sure your **attic fan** is ventilated. An attic fan will remove hot air during warm weather to lower air-conditioner operating time, or, perhaps, to avoid the need for air-conditioning altogether.

Be sure to check the **fresh-air intakes** under the eaves of your home. If they are covered with screens, clean them to ensure proper ventilation.

In homes with air-conditioning, ventilating fans should be turned off promptly when their job is done to avoid exhausting cooled air to the outside.

Average installed **cost** of an appropriate powered attic-space ventilator runs around $200, depending on the area to be ventilated and the installation requirements. It pays to start with a reliable dealer who can recommend the proper air movement capacity for the area to be ventilated. Get two or three bids on the job before making a final decision.

Also, look for the *Home Ventilating Institute certification* label. This certifies ventilating performance in terms of air movement in cubic feet per minute (cfm) and the sound output for room ventilators. The Institute has also established standards for desirable air changes per hour for different areas—with the cfm rating required to meet these standards. Two publications, "Certified Home Ventilating Products Directory," and "Home Ventilation Guide," should be available from any dealer who sells HVI-certified products, or they may be purchased (50 cents each) from Home Ventilating Institute, 230 North Michigan Avenue, Chicago, Illinois 60601.

Other Types of Fans

When natural air circulation is inadequate, use a **portable fan.** Place it by a window to draw in air, and position the fan direction so there is no uncomfortable draft on you.

A window fan or a "**whole-house ventilating fan**" will do a bigger job. The hot spot of your house is, of course, your attic. Therefore, an attic fan or a whole-house fan is a very economical investment, especially if you can't insulate the roof. Even if you have an air-conditioner, use the fan at night to draw in cool air when the air-conditioner is off.

The attic fan should be positioned on one side of the attic and an open louvered vent positioned on the other side so that cool air can be drawn in. When the outdoor air is cooler than the indoor air (and the air-conditioner is off), leave the attic door open so air within the whole house can circulate.

If you must close the house for a period of time, an automatic temperature-set thermostat can operate the fan even while you are away.

CHAPTER 5

Don't Burn Money to Keep Warm

HOMEOWNERS and apartment dwellers can save fuel and keep heating bills under control this winter by taking many measures now to keep heat in and cold air out.

Insulation

In most homes, the big heat sieve is the **roof,** which should be backed with at least six inches of insulation. Depending upon your area and the size of your roof, insulation will cost between $100 and $300, and this outlay will be easily recouped in one or two winters of lower heating bills.

It's easy to check the effectiveness of insulation between the **walls** in cold weather. Go to a corner of a room, place one hand on a wall that is a partition between two rooms, and the other hand against an outside wall. If the outside wall feels much colder, its insulation is poor.

Some companies now blow granular or **fibrous insulation** into wall cavities. Before you agree to this treatment, make certain that the construction of your house lends itself to this work, and that the insulation would not create condensation inside the walls. Have the construction of your house evaluated by a legitimate firm before contemplating blown insulation.

Adequate insulation and weatherstripping alone can cut annual heating costs by as much as 50 percent. Older homes especially should be carefully checked. If additional insulation is needed, it can be a do-it-yourself project, or turned over to a reputable insulating or building contractor. Expenses pay for themselves in a few years through reduced heating costs.

Weatherstripping

How can you **check** your weatherstripping? On a chilly day place your hand at the frames of doors and windows. Feel any drafts? They mean air leakage. Sometimes old weatherstripping may be worn or torn and need replacing. Sometimes doors and windows warp or shrink, making weatherstripping ineffective.

Felt, rubber and plastic weatherstripping, usually with a peel-off adhesive backing, can be used around door and window frames. Metal plates with a rubber or vinyl insert can be screwed to the door's threshold; or a rubber or felt strip can be attached near the bottom of the door to seal out drafts and keep in the heat. (For more detail, see Chapter 11.)

Caulk cracks and seams on the exterior of the house. Even a hairline crack or seam can allow a lot of cold air to come inside. Fill these openings with caulking compound. If old caulking has dried out and shrunk so there are spaces between it and the wood, scrape it out and replace it with fresh caulking. Typical spots that need caulking are seams where siding joins window and door frames, and corner boards; also along a cornice where siding joins the roof. Also caulk around any fitting set into the outside wall, such as electric lamps and exhaust-fan outlets. The easiest way to apply caulking compound is to use a caulking cartridge in a caulking gun. The cost of a cartridge of caulk will run from around 69 cents to over $4.99. The more costly compounds last longer than the low-cost, oil-base compounds, which may last only a year or two. You can figure that one cartridge will produce a one-fourth-inch-wide bead around twenty-five feet long.

Fill cracks in foundation walls. Any above ground level will let in cold air. Seal large cracks with cement mortar, hairline cracks with caulking compound. Be sure to seal the joint between the top of the foundation wall and the house sill. The seam can be a big source of heat-loss.

Any loose caulk around **doors and windows** should be

removed with a screwdriver (use like you would a knife) and replaced. Check the rubber sweep plate at the bottom of your storm door. If it is not touching the threshold, the threshold should be raised with a screwdriver and the space filled in with caulk.

It costs about $250 a year to heat a small well-sealed home in the Midwest, but an unprepared home can raise the cost of heating 25 percent or more. **Exterior joints** where leaks can occur, such as those between the porch and house, should also be checked and sealed. **Roof leaks** can cause a tremendous amount of damage in a short time. Roof sealer is also available at hardware stores and lumber yards, and should be daubed on with a stick or heavy paint brush over cracks and gaps.

Storm Windows and Doors

Storm doors and windows **pay for themselves** very shortly, especially if installed in northern and western walls. Double glass can cut down heat-loss through a window by as much as 50 percent, and this factor can reduce your total heat bill by roughly 10 percent.

Unless you don't plan to leave the house all winter, you are going to be **opening and closing the doors,** which results in an inevitable heat-loss. Even if you don't open the doors, if the weatherstripping is defective, you will lose heat anyway. Hence, it's a good idea to check all weatherstripping and replace it if worn. Believe it or not, an active child, running in and out of the house, can tack 3 percent onto the cost of heat for the winter.

Air Supply for Your Furnace

During combustion, your furnace gulps a huge quantity of indoor air and sends it billowing up the chimney (an average of three normal-sized bedrooms full of air in an hour's combustion.) If it lacks its own outside air source, and it commonly does, the furnace will suck heated air from the rest of the house into the cellar. In a tight house it would probably not get enough for proper combustion, and up to half the fuel may be wasted. The cure: give the furnace its own air supply by **opening a basement window** a few inches or by having an air duct to it installed from outside.

Humidity

Cold air can hold less moisture than warm air. For this reason, old, leaky houses are likely to be dry. To **raise humidity,** tighten up the house with weatherstripping and insulation. If necessary, buy a humidifier.

New, tightly built, small houses tend to be too humid; water runs down the windows, the doors stick and, if there is no proper moisture barrier in the walls, there may be damage to the house. To **lower humidity,** try ventilation. Install ventilating fans in the kitchen and bathrooms, and give the clothes dryer an outdoor vent. If necessary, invest a hundred dollars or so in a mechanical dehumidifier.

Humidifiers may be built on your furnace, or may be separate units. They make the air feel warmer so you can be comfortable with less heat. A **dehumidifier** can be just as helpful in hot weather, maintaining comfort at higher temperature levels. (See Chapter 9 for details on purchasing and using humidifiers and dehumidifiers.)

Your Thermostat Setting

Conserving the amount of heat you use at home has always made sense, but this year there are strong, new incentives to actually cut down. As the cost of heating fuel continues to rise, every gallon of oil, therm of gas or kilowatt-hour of electricity conserved adds up to more and more money saved. At the same time, the unused fuel contributes to national energy conservation, and also lessens the possibility of local shortages, such as occur quite often in many parts of the country.

Since the amount of heat you need depends directly on the indoor-outdoor temperature differential, **the lower you can comfortably set your thermostat, the less fuel you will use.** Consider whether you have been overheating your house. The sense of comfort gained from being warmed or cooled is mostly a matter of conditioning. Nearly everyone has had the experience of visiting a home that seemed uncomfortably hot (or cold) except to those who lived in it. Obviously some people feel comfortable in temperatures that are higher than average; others prefer lower temperatures. Since it is possible to save about 3 percent of your heating expenses with every degree above 72 degrees Fahrenheit that you lower your thermostat, it would certainly pay to experiment—and see if you and your family can adjust to somewhat cooler indoor temperatures in the colder months. Heavier clothing will help, though we wouldn't want anyone sitting around the house in an overcoat (a light sweater would be a good compromise.) This is the easiest cost-nothing step you can take, and whether or not you lower your thermostat setting in the daytime, you should certainly do so at bedtime. This procedure, once considered of doubtful value, was recently tested under carefully controlled conditions and proved to conserve energy. To keep warmer on chilly nights, place a lightweight **blanket** *under* your fitted bedsheet.

The actual amount of fuel you can save depends on several factors—where you live, how much you lower the thermostat and for how long. Honeywell, Inc., a leading maker of heating controls, has computer calculations that may give you a general idea. They show that if your normal daytime thermostat setting is 75 degrees Fahrenheit, turning it down to 67.5 degrees for eight hours during the night reduces fuel use 8 percent in Milwaukee and Buffalo; 9 percent in Boston, Chicago and Denver; 10 percent in New York, Cleveland, St. Louis and Seattle; 11 percent in Louisville, Washington, D.C. and Portland, Oregon; 12 percent in San Francisco; 13 percent in Atlanta and Dallas; 14 percent in Los Angeles.

And the Bureau of Mines further estimated that American homeowners, through proper conservation measures, could cut their fuel costs by as much as $1 billion a year by turning thermostats down. For comfortable and economical furnace operation, maintain an average household temperature of **68 degrees** instead of the nearly 75 that is customary in most homes. Government experts say that this procedure alone will cut about 10 percent off your fuel bill.

A recent study completed at the Holifield National Laboratory in Oak Ridge, Tennessee, echoes this claim, noting that if all homes in the United States would reduce heat from 72 to 68 degrees in the daytime and to **55 degrees at night,** our national energy budget could be cut by 4 percent. The saving is equal to 25 percent of our petroleum imports.

Don't be a thermostat fiddler. Using fuel at an even pace helps conserve it, so **avoid constantly turning the heat up or down.** If operating properly, your furnace will automatically turn itself off and on to maintain the temperature you have set. Turning the thermostat up high doesn't make the house heat up any faster, since the furnace air blower runs at a constant speed.

And, many doctors say the cooler temperatures are healthier. Especially for older people, heat rather than cold, puts a strain on the heart. Put **warm coverings on the floor** so your feet won't feel cold. And live with **plants** to raise the humidity and make cooler temperatures more comfortable. It will also help to **dress in layers**—which is very fashionable today, in addition to being comfortable.

Statistics show annual heating costs rise three to four percent for each degree above 68 degrees. For example, if your annual bill is $234 at a constant 68 degrees, you could boost it to a season total of $320 by setting it at 76 degrees. Many doctors contend this is not a good idea.

Greater savings can be realized if you lower that thermostat **10 degrees,** but don't go any further or you will likely waste more fuel raising the house temperature to the desired level in the morning than you've saved by lowering it at night. Furthermore, you may feel pretty uncomfortable during that extra-long warm-up period. To maximize your savings, the setback should be done regularly every night, regardless of the outside temperature. A clock-controlled thermostat takes care of this automatically, at the time and temperature you select, and it will also turn up the heat every morning. Installing a clock-thermostat will cost between $50 and $90, but it ensures your setback every night, and you will recover the cost after several heating seasons.

If you leave your house for an extended period during the cold months, set the thermostat as low as is safe in your area. Most go down to 55 degrees—generally considered a good lower limit for safeguarding your pipes.

Care of Your Heating Plant

Clean your heating and cooling filters at least once a month. Replace worn out filters regularly; if you can't see through them, they are ready for pitching. A dirty filter restricts air-flow and makes the furnace run longer to achieve a given temperature.

Vacuum **floor grilles** or registers thoroughly.

If you open windows or use a fireplace in winter, **shut doors** leading to other parts of the house. If you don't, warm air from the rest of the house will leak out through the open window or be drawn off through the fireplace. Always keep the **damper** closed when the fireplace isn't being used. A good chimney can draw off up to 20 percent of the heated air each hour it is not blocked.

When your fireplace doesn't have a damper which you can close when it is not in use, **block the opening** with plywood or insulating board (or a piece of metal-backed asbestos.) An open chimney "draws"; if it isn't drawing smoke from burning wood, it is drawing air warmed by your furnace—and money out of your pocket.

Clean all **radiators and vents.** Be sure that rugs and furniture do not obstruct floor or wall outlets and that radiator valves have been opened.

Keep the **thermostat box** free of dust, and be sure the air circulates freely through it.

Inspect warm-air **heating ducts** for holes. Holes, cracks and open joints in metal heating ducts waste heat. Seal them with aluminum-foil duct tape. A sixty-foot roll 2½ inches wide costs from $4 up. Duct tape is useful for sealing joints around metal frames of windows that don't have to be opened in winter.

Keep heat outlets, especially forced-air ducts, free of obstructions.

You can increase efficiency of your radiators by taping **aluminum foil** on the wall behind the radiator. The foil will reflect radiated warmth back into the room and will keep a cold outside wall from cooling the radiator.

Vacuum-clean radiators, baseboard units and convectors periodically. Even a thin coat of dust on the heating surface acts as insulation and reduces heat output, thereby wasting fuel. Radiator covers also reduce heat output, so it is wise to remove them during the heating season. If you can't stand the sight of an ugly radiator, paint it the same color as the wall to make it less conspicuous. However, do not use a metallic paint, such as aluminum or bronze, on radiators. It reduces heat output. Use a flat paint or radiator enamel. Never paint radiator *valves*, though. You can also replace the boxed radiator top with grill work to add a bit of beauty to the room.

Furnace tune-ups Be sure your furnace is "tuned." It should be **serviced** once a year, preferably in the summer or early fall. Don't wait until the first cool days arrive, since that is when heating contractors and gas company service personnel are swamped with calls. A properly adjusted furnace could save up to 10 percent in fuel consumption during a heating season. Regular checkups make sure that it is operating properly and at peak efficiency—and thus using a minimum of energy to do its job.

When the professional looks over your furnace he will check the burner adjustment, clean the blowers, change the filters and scrutinize your thermostat to make sure it is not being misled by dust or lint. In these short times, consider adding the new "fuel sentry system" to your thermostat. It can be programmed to suit your schedule, automatically lowering the temperature at bedtime and raising it before you rise.

Remember, old or new, a furnace that burns oil or gas can become **inefficient.** This happens when products of burning collect on surfaces meant to transfer heat or when too much or too little air enters the combustion chamber. In an oil-burning furnace visible smoke is a sign of low efficiency. In a gas burner a flame with too much orange or yellow indicates a problem. And any forced-air unit with a blower motor that is switching off and on too frequently isn't functioning efficiently.

Only a competent serviceman can tell you for sure whether your furnace is working efficiently. His examination takes a few minutes, for a test that records carbon dioxide and smoke content and the temperature of combustion gases going up the flue.

If you have oil heat, be sure to check to see if the **firing rate** is correct. Chances are it isn't. A recent survey found that 97 percent of the furnaces checked were overfired.

In an oil furnace the burner should also be adjusted for top performance (or, if necessary, replaced), and soot and scale removed from inside walls of the **firebox.** These simple measures can increase burner efficiency by 10 to 12 percent. Improvements up to 25 percent are possible with installation of reflective insulation in the combustion chamber. Most oil heaters built in the last ten years have it, but not those installed before.

Also, if your furnace burns oil, ask the fuel supplier whether **additives** have been put in the fuel to keep the chimney soot at a minimum.

Gas burners also get out of kilter, and having the **carbon dioxide test** done on them is a good idea. Once adjusted, a gas furnace usually requires less attention than an oil-burner, although it should be checked and cleaned every year to maintain peak performance. Some gas companies inspect and adjust furnaces at no cost, others charge a nominal sum, and still others contract services out to private firms.

Hot water systems should be cleaned of scale and rust from the combustion chamber. The circulator pump should be oiled sufficiently. Keep water at the proper height or pressure as indicated by the boiler gauge. Every so often, open the radiator vent valve slightly to let air out. If the radiator becomes air bound, efficiency is reduced.

Buying a New Furnace

If you are getting a new furnace, make sure it is the **right size** for your needs, and avoid wasting energy by buying any unit that is larger than you need.

If you need a new furnace, buy one that includes an **automatic flue gas damper,** which reduces heat-loss when your furnace is not in operation.

And, if you are considering electric heating, consider a **"heat pump" system,** which uses outside air in both heating and cooling; it can slash your use of electricity for heating by 60 percent or more (see Chapter 8).

Degree days   While you are waiting for the serviceman to come, a little mathematical exercise may prove useful and enlightening. Ask the nearest U.S. Weather Bureau how many degree-days it recorded in each of the preceding three or four years. (A degree-day is simply a unit of temperature used to measure the amount of heat, and thus the amount of fuel, needed to maintain a given indoor temperature in a given region). And, consult your fuel supplier or your own records to find out how many gallons of oil, therms of gas or kilowatt-hours of electricity you bought in each of the same years. Then divide the number of degree-days for each year by the corresponding fuel-consumption figure. Compare the results for the three or four years. If there is no more than a 5-percent difference between them, your heating system is in good operating condition. But, if you've used considerably more fuel per degree-day in the last year or two, something may have happened to your furnace. Unless there is a simple explanation—a change in your living pattern, for example—chances are the heating system has gone awry. Tell the serviceman to look for serious trouble.

Planning for Retaining Heat

And, be sure to **close all doors** to the attic, basement, garage or other unheated parts of the house so that you won't spend money heating or cooling spaces when they are not in use.

It's best to remove air-conditioning **window units** in winter. But if you keep them in, cover them tightly.

Try to recycle some heat. For instance, if you have an **electric clothes dryer,** you can pull the vent back into the house, put a stocking over the end to catch the lint, plug up the outside hole and let the dryer warm up the room while it is drying the clothes. (However, this may prolong the time your dryer takes to dry your clothes, as noted in Chapter 3.)

Use **solar heat** whenever possible. When the sun is shining, open window coverings and let Old Sol warm your house as if it were a greenhouse. Other times, keep window coverings closed so they can act as insulation against the cold air outside.

Unless you have an outside source of air for your furnace, don't seal your house too tightly. Your gas appliances make use of the normal seepage of **fresh air** into a house as their source of oxygen.

Never use a gas range or oven to heat a kitchen. For instance, when the oven door is left open to heat a room, the temperature keeps climbing, which could cause a fire such as the oven's insulation burning.

Zoned Heating

Dividing a house into heat zones can save energy, especially in a new home, or, where practical, in an older home. Four zones usually do the job nicely—north, south, upstairs and downstairs. For instance, why pour heat into the sunny south side of the house just to satisfy the heating needs of the shaded north side? Put **automatic damper controls** into the ductwork in the basement. It is also possible to put dampers behind the outlet grilles in the several rooms. All these controls can be put on a time clock (about $30 each) to control the various thermostats.

Here is how the zone system works. Let's say nobody is in the living areas from 11 P.M. until 7 A.M. And, in the daytime the room is occupied. Further, in the daytime the north side of the house needs more heat than the south side. With automatic controls, the dampers will respond to the thermostats in the four zones; each thermostat will call for whatever temperature you have set, and "demand" that the duct deliver enough heat to satisfy the setting. And, if that same thermostat is on a time clock, it will call for certain degrees at certain times of the day and night. Cooling, once again, is just the reverse. The warmer parts of the home can be programmed to receive more cool air than the shaded areas. And, there is no reason why you should want to cool any part of the house that is not in use. Zoned heating and cooling definitely offers flexibility as well as economy.

Installation of zoned heating is a bit more complicated in a split-level house, though certainly worthwhile in the long run.

Auxiliary Heating Units

We are told that permanently installed auxiliary heating units institute one of the fastest-growing segments of the home appliance industry. These are handy to have in your garage on a cold winter day, or perhaps in a recreation room that isn't used every day.

The demand for these secondary home heating sources is greatest in the South where only occasional bouts of cold weather are experienced, and in the North where bitter cold weather is sometimes too much for the central heating system to handle. The units are found in home areas where almost instant warmth is desired, where it isn't necessary for the thermostat to bring up the heat, and where there is no need to spread the heat throughout the entire house: in the bathroom, to ensure quick warmth when bathing or showering; in basements that are used occasionally as recreation areas; in workshops and laundry rooms adjacent to unheated garages.

Historically, provision for auxiliary heating began to appear in the better-built homes in the South and Southeast about 1949 in places where spring and fall often bring warm days but chilly nights. Builders found that homeowners preferred this type of secondary heating to starting up the central heating system every cool evening.

There are three principal types of auxiliary units now on the market: **electric heaters,** ranging in heat output from 250 watts to 1800 watts, in small increments; units producing **infra-red radiant heat;** and **fan-forced warm air heat.** Heaters come equipped for wall or ceiling installations. Most builders prefer to place these units in the ceiling, especially in homes that are to be sold to families with small children. These ceiling models are made for flush or surface mounting. Some ceiling units also have exhaust fans and lights.

Portable electric heaters, too, may be economical when used for extra heat in small areas. But be sure that any you use are thermostatically controlled. If not, one average-sized electric heater, operated twelve hours a day, could add a great deal to your monthly utility bill.

As another idea, (Grandma would love this), consider a **woodburning unit** you can use alone in cabins, garages, etc., or piggyback to your present gas or oil-burning furnace to relieve it of some of its work load. The manufacturers say you only add wood (or household burnable trash) every six to eight hours with some stoves to keep the energy-converter piping along. Where wood is readily available and winters are cold, many families depend on a wood stove for much if not all of their heat. A wood stove is many times more efficient than the conventional fireplace. It can be set in front of a fireplace and connected with the chimney, or it can be provided with its own prefab metal chimney. Some of the most efficient wood stoves are imported from Scandinavia. Wood stoves can cost from under $100 to over $1000, depending on size and efficiency.

CHAPTER **6**

Help Heat Your Home with Firewood

THERE'S SOMETHING about a fireplace that turns a house into a home. However, if you install a fireplace and only burn a fire in it on Saturday nights, it isn't going to put much of a dent in your heating bill. Using a fireplace or a wood-burning stove **consistently** throughout the winter will help. We must all agree that heating our homes with fuels such as natural gas, coal and oil is getting more and more expensive. Not only that, we may soon reach the point where those sources are simply no longer available, at any price. Where firewood is available, it may offer a reliable and relatively low-cost alternative to those prepared to use it for heat.

David Dyer, a mechanical engineering professor at Auburn University, said in 1979, "With heating oil going to 90 cents or $1 or more a gallon this winter, you will see more interest going back to wood, because people just can't afford oil." Dyer estimated some homeowners in the Northeast can expect to pay up to $3000 for heating fuel this winter. He adds, "People who are running on natural gas right now are going to be in the same situation as people heating with oil (because of gas price deregulation). But, people have access to wood, and really it is the only way out. People have been using wood for centuries, but nobody knew much about it." But, before buying a wood-burning fireplace or stove, Dyer suggests, consumers should make sure of what they are getting.

Buying a Stove

When choosing a wood stove or fireplace, remember that **bigger isn't always better.** You should choose a model with a heating capacity that meets, but doesn't exceed, your home's heating needs. Most manufacturers list the BTU output on their fireplaces or heating stoves. But it would be well to keep in mind that how well a fireplace or stove works in your home depends not only on the BTU output, but also on the type of wood you decide to burn, the outside temperature, the frequency of use, and, of course, the ability of your home to hold the heat that is produced.

Wood-Burning Stoves

One of the hottest conservation devices of the 1980's is the wood-burning stove. Whether your goal is saving money on fuel, keeping your home toasty warm or beautifying a room, Americans are expected to spend more than $1 billion on these devices by the end of the year.

But, before you join the rush to spend from $300 to $5,000 for a wood-heating unit, be warned. Here are some guidelines, from the experts—heating consultants, fire marshals and makers of wood-burning stoves and furnaces:

• Be sure the equipment you buy has been certified by an established testing organization—for instance, Underwriters Laboratories (UL)—or approved by your local building inspector.

• Some wood-burning stoves and furnaces are more efficient than others, but to date no reliable testing agency has compared those now being sold on the market. Larry Gay, founder of a firm that makes both wood stoves and furnaces, says that BTU ratings "assigned to wood furnaces by manufacturers are crude estimates at best and not even based on a standard set of assumptions."

• Heating with wood is more dangerous than heating with gas, oil or electricity. Some units tend to have fewer built-in automatic safety devices. However, the major problems are primarily improper installation, operation and maintenance rather than poor equipment itself. Have your local fire department or marshal inspect your system after installation to be sure the work has been done correctly from start to finish.

• Never use gasoline or kerosene or fake logs to start or rekindle a fire. They can cause an explosion or harm your unit.

• Don't burn wood that is green or damp. This encourages the buildup of creosote—a thick, tar-like and highly flammable substance that lines stove pipes, flues and chimneys if they are not cleaned properly.

• Install smoke detectors on each floor of your home, especially outside bedrooms. The danger of being fatally overcome by smoke is more likely if you are heating with wood than if you are using another heating system.

Fireplaces

Fireplace operation   Energy experts also tell us **not to use a fireplace for supplemental heating when your furnace is on.** Why? Because, as suggested in the preceding chapter, warmth from a fire on the hearth doesn't usually radiate through the entire house. The heat-gain is confined to the one room with the fireplace. And, when your furnace is on, too, a considerable amount of heated air from the rest of the house flows into the fireplace and goes wastefully up the chimney. Then the temperature in other rooms of the house goes down, and the furnace uses more fuel to raise it to the level controlled by the thermostat. So you use more fuel, rather than less, when the furnace and the fireplace are both going.

If you must use your fireplace when the furnace is on you can **lessen heat loss** by **lowering the thermostat** setting to 50 to 55 degrees. Some warmed air will be lost, but the furnace won't have to use as much fuel to heat the rest of the house as it would to raise the heat to 65 or 68 degrees.

Close all doors and warm air ducts entering the room with the fireplace and **open a window** near the fireplace about one-half to one inch. Air needed by the fire will be provided through the open window, and the amount of heated air drawn from the rest of the house will be reduced.

If you have a simple open masonry fireplace, consider installing a **glass front** or glass screen. This will cut down on the loss of warmed air through the flue.

Pre-fabricated models   If you are planning to add a fireplace to your home, be sure to consider the various **prefabricated models.** They are more energy-efficient than regular fireplaces and can be used as supplementary heating devices.

The **Franklin Fireplace/Stove** is fast becoming one of the more popular models. Because it is usually made of cast iron and set out in the room, it will radiate heat from all sides while holding the heat very well. To use a Franklin stove as a fireplace, just open the doors. When they are closed, the stove acts as an effective room heater. And, these units will give off a great deal of heat from a relatively small amount of wood. Some models also have special systems that circulate the warmed air in order to keep the room temperature more even.

Another choice is a prefabricated wood-burning metal unit that is set in place, brick or stone then being put around it to give it a built-in look. These units are also available with **heat-circulating systems** that draw air in from the room, circulate it around an inner shell in the firebox, warm it and then expel it back into all corners of the room. Models with blower fans increase the flow of room air throughout the circulation system and heat the room faster.

Some of these units can be used with ductwork and heat registers so that the warm air can be diverted to adjoining or upstairs rooms. Some of the very newest models have even more advanced energy-saving features for you. One model is designed to take the air needed for combustion from outside the house, rather than burn up valuable heated room air. This unit also has the air-circulating feature. It comes with a glass fire screen that prevents the escape of warm air up the chimney while radiating additional heat. Because of its special design, the unit gives more room heat per log burned. With any prefabricated fireplace, be sure it is Underwriters Laboratories listed, and has their seal of approval to assure a quality product.

New devices for Increasing Heat from Fireplaces   There are many devices on the market today that you can add to any conventional fireplace to increase its heat efficiency as well as output. (*Heating efficiency* is the net amount of heat gained for the amount of fuel burned. *Heat output* is simply the amount of heat produced by a heat source at a given time, regardless of the amount of fuel consumed.) Most of these devices will reflect most of the heat back into the house. For instance, one new unit draws cool air from the outside, warms and circulates it into the room. Another hooks up to your existing boiler so it can help heat the water that goes on to warm the rest of the house. Others incorporate glass doors to conserve the room's warm air, or electric blowers to distribute the fireplace-heated air evenly through the house.

The new energy-saving **fire screens** are great. You can't shut your damper if a fire is still burning when you go to go out or turn in. But you can cut your heat loss with a glass fire screen ($125 and up.) These fire screens have heat-resistant-glass folding doors you can close to keep the heat inside your house from going up the chimney with the smoke. (The experts claim a roaring fire can actually chill your home because it sends nearly 12,000 cubic feet of warm air up your chimney every hour! And, as we have suggested in Chapter 5, this includes air you have paid your furnace to heat.)

No matter what kind of screen you choose, make sure you have a broad, non-tippable screen to hold back sparks and flying cinders.

Grate-type blowers are also used with fireplaces to increase heat production. You build your fire directly on top of the unit, as you would with an ordinary grate. Heat from the coals is then transferred to the interior of the grate, where incoming air is warmed as it passes through a series of baffles. A blower then forces heated air through tubes at floor level, chasing away the layer of cold air that usually accumulates next to the floor.

This type of unit can increase your heat production by 500 to 600 percent, and because so little fuel is required, the unit can increase your fireplace efficiency by 900 to 1,000 percent. The grate-type blower unit is stingy with fuel and does not require a roaring fire to produce additional heat. Actually, even after the fire has burned down to a bed of coals, the device will generate heat for another three or four hours by pulling heat out of the coals. (Grate-type blowers are priced under $200 to $250 in most parts of the country.)

Nearly all the new fireplace add-on heating devices work by heating air, then conveying it directly into the room. The most common item on the market for this purpose is the **curved-tube convection grate**. All models are similar in design and function, and all utilize the same heating principle. And, best of all, they work. You will find approximately thirty manufacturers produce units of this type.

In essence, the curved grate is quite simple. It consists of a set of tubes in a C-shape, placed in the fireplace with the open side of the C toward the room. You build your fire over lower tubes. The molecules of heated air in the tubes then rise, causing an upward flow of air through the tubes. The rising heated air is then forced out of the tubes at the upper end, and new air is sucked into the bottom tubes in a continuous flow.

The simplest and most basic kind of tube heater has no blower to force air through the tubes. To blow warmed air clear of the fireplace opening, the heater depends solely on the rising and expanding movements of the heated air. If you have a non-blower model, be sure to position it with the upper ends of the tubes within two inches of the top of the fireplace opening. This placement will help get the heat far enough into the room for the heat to rise away from the opening and not be dragged back into the fireplace with air needed for combustion.

A curved-tube convection grate without a blower costs between $50 and $100 and can increase your heat output of the fireplace about 50 percent. However, there are a few disadvantages. Ash removal can be a problem when the unit is used along with a glass door enclosure. Most fireplaces are not deep enough to allow you to slide the tube unit back far enough to reach the ashes easily. Also, if you have a conventional damper control, it might be difficult to operate if there is insufficient space between the tubes to let you reach the damper control arm. However, you can easily get around this problem by fashioning a hooked rod to reach through the tubes.

Blower-equipped tube grates are equipped with an electric fan and motor attachment which propels heated air beyond the fireplace opening and well into the room. These range in price from about $125 to $200. They also require a nearby electrical outlet.

Some fireplace devices combine a curved-tube grate with **a glass fireplace enclosure.** With such a unit, you can burn a fire with the glass doors closed and the air circulates through tubes in the door frames and into the room. With a normal fire, a glass-doored convector can increase the heat output of a fireplace by about 300 percent, because of the combined effect of the heating tubes plus the heat that radiates from the glass doors. The glass doors also keep heated room air from escaping up the chimney. Only the air that is needed for combustion is allowed to enter the fire. (Glass-doored convectors cost between $300 and $450 depending on the region in which you live and competition between shops handling this item.)

Check your fireplace specialty shop or local home center. Fireplace adaptors to increase heat sell for as little as $50 (or as much as $800), and you can make your fireplace as practical as it is cozy and romantic.

Firewood Buying Tips

Unless you have a supply of free wood, it's going to cost you to burn a fire. You have to compare the cost and BTU output for wood burned in a fireplace or stove with the cost and BTU output for the source of heat you already have.

Always buy **full cords.** Specify standard cord when you order, and make sure that you get a pile of wood that is four feet high, eight feet wide and four feet long. Sometimes you will hear the term "face cords"—stacks eight feet wide and four feet high, but cut in less than four-foot lengths. "Short cord" is another name for face cord. Shop around for your wood, and never pay full-cord price for a face cord.

Buy your fireplace wood in quantity. A "rick" (one-third

of a cord) or a pickup truckload delivered when needed throughout the season usually ends up costing more than one delivery of several cords.

Measure the woodpile to be sure it is **tightly stacked** and all the pieces lie in the same direction. Otherwise you will pay for air-space instead of wood. Even piled carefully, a cord contains only 80 cubic feet of solid wood, although its dimensions total 128 cubic feet.

Size and straightness of wood in a pile also help determine how much wood you get. Crooked pieces add to air-space, but large pieces add up to more wood, even though the spaces may look larger.

Know **what kind of wood** you are buying. Firewood dealers sometimes advertise "mixed hardwoods." Since all hardwoods are not equally desirable as firewood, find out what the mix includes. Buy only the woods that suit your heating needs. Use seasoned hardwood logs, such as oak, ash, maple, or hickory. Green logs will give off smoke and are difficult to ignite.

Pound for pound, most species of dry wood produce about the same amount of heat. But some species of wood are lighter than others. For instance, a cord of aspen weighs about half as much as a cord of dry white oak and has about half the heat value. If you live in an area where both aspen and white oak are available, the white oak would be the better buy. As a general rule of thumb, the heavier woods are the greatest heat-producers.

If you are going to use the wood immediately, never buy **green wood.** Wood suitable for burning has ideally been air-dried for a year and contains about 20 percent moisture. Never burn wood with a higher moisture content. Green wood is not only inefficient, but it is a dangerous fuel, because its extra moisture allows a rapid buildup of creosote in the chimney, which is the cause of most house fires connected with wood heating.

There is a way to save on labor costs when buying wood. Trees are normally sawed into **8-foot** "sticks" for easier handling and hauling. The dealer then cuts the sticks into smaller lengths. A large-volume wood dealer will often sell you these sticks at a considerable discount. Then you can cut the wood to suitable lengths at your leisure.

Plan ahead when you buy your firewood. Check with local

wood suppliers about purchasing your winter firewood supply a year ahead of time. If you have enough room outside to air-dry newly cut green wood for twelve months, you can save as much as 50 percent.

Miscellaneous tips Popular Mechanics Encyclopedia says, "Though there are times when you may use charcoal, coal or commercial pressed logs, it is pretty safe to bet that you, like most, actually prefer to use real wood." Thus, with the cost of firewood as it is today you must know what to select, how to cut and stack logs as well as how to use them properly in the fireplace.

Remember, softwoods like pine, fir and spruce are easy to ignite and burn very quickly with a hot flame. The very speed with which softwoods burn makes them less desirable. Also, burning softwood over a period of time can build up a dangerous coating of tars in your chimney.

You should also avoid using scrap lumber and refuse. These materials, especially when excessively dry, produce sparks which escape up the flue and are a dangerous fire hazard.

To obtain the best fire—one that burns long and gives plenty of heat—combine softwoods with hardwoods such as oak, birch, maple and ash. The hardwood species burn less vigorously and with a shorter flame—they burn more slowly. For a pleasant aroma, add woods from fruit and nut trees—apple, cherry, pecan and hickory—which all give pleasant scents. Generally, wood smoke's scent resembles the fragrance of the tree's fruit.

Position the longest and thickest log across the andirons toward the rear of the fireplace. Do not let the logs butt tightly against the back wall. Leave about ½ inch of space between them. Place a second piece, preferably a split log, in front, and a third split log on top to form a triangular pile.

If you use a fire basket instead of andirons, the fire-laying procedure is the same: just place crumpled newspaper below the basket.

Don't clean out your fireplace after each fire. Ashes about one-inch thick should be left and spread evenly over the entire hearth. These will insulate the cold hearth and make the next fire easier to start.

Don't build too big a fire. Three or four logs will usually provide all the heat and flames you will want. As top and front logs burn up, turn them with tongs and, if necessary, place another split log on top.

Never buy wood by weight. Wood is sometimes advertised for sale by the ton. But the various types of wood have different densities and moisture contents, and the weight difference can be very significant. The water can account for as much as 25 percent or more of the price of the wood. One cord of wood produces as much heat as 146 gallons of heating oil, 17,400 cubic feet of gas, 3,300 kilowatt hours of electricity or just under a ton of coal.

Sources of cheap firewood   Keeping up a good, roaring fire can supplement your home heating and help you save energy costs. But, fires need replenishing and wood can be expensive. Here are some ways to get firewood free, or for a small fee:

1. We have 164 **national forests** that allow a certain amount of tree-chopping. You can get permission to chop live trees or collect dead ones from the administrator of a forest. However, remember that national *parks* prohibit chopping of trees.

When you do get permission to go into a forest, try to select **a blend of hardwood and softwood** trees. Hardwoods burn longer and so have more heat value; softwoods are perfect for short early-morning fires to drive out overnight dampness and cold. And don't forget the aromatic woods, such as those of fruit trees. They add a pleasant dimension to your fire. For 25 cents you can receive a copy of a chart that rates firewood. Write to the Superintendent of Documents, U.S. Government Printing Office, Washington, D.C. 20402. Ask for "Firewood for Your Fireplace" Publication Number 559.

2. Many of our **state, county or city parks** have trees for the taking. Again, check with the proper authorities. District foresters would be in charge of state forests, extension agents for county forests, and city foresters for municipal lands.

3. Municipal **park and highway departments** chop down and store many trees. Check with municipal officials. You may be able to chop up and haul away some of these trees.

4. Maintenance departments of local **power companies**

oftentimes chop down branches and trees that are too close to power lines. They would be pleased to thin out some of their accumulation. All you need do is call your local customer service representative and ask. However, do not take any old power line poles. They are dangerous when used as firewood because they are so heavily creosoted.

5. **Construction or remodeling sites** also generate piles of dry wood cut into pieces suitable for burning. Contractors have to pay someone to haul waste wood away, so they would be happy to give it away or sell it cheap. Don't forget about buildings under demolition. They often contain lots of dry wood.

6. **Trash dumps** nearly always contain fallen limbs and dead trees, if your local regulations prohibit open burning. And, if the sanitation workers or contract carters must separate trees and branches at the dump, it will be easy to collect them there. Get permission first.

7. **Firms that work with wood**—furniture manufacturers, lumberyards, carpentry shops—usually have to pay someone to haul away scrap. They are often willing to give it away or sell it for a small price.

Store your firewood outdoors under some kind of shelter to protect it from rain and snow. You can make a "roof" over the wood pile out of heavy plastic sheeting held down by a few logs. Wood dries out faster in small pieces, so split it before stacking.

Building Your Fire

The fireplace or stove **damper** should be open all the way before lighting a fire and, in most fireplaces, should remain open wide at all times when there is a fire burning. Otherwise smoke will back up into the room.

Before making the fire, be sure there is a good **draft**. This is essential to pull in the air so that the fire will burn properly, as well as to carry the smoke up the chimney. It may be necessary to open a window or door slightly to supply the fire with air.

But, an open staircase in the room or a floor register installed in front of the hearth will do the job.

Lighted **stubs of candles** placed under kindling make good fire starters. Use long fireplace matches or twisted newspaper to light the crumpled newspapers and kindling wood. Some fire-makers also use the initial torch to warm the flue as a protection against smoking.

Never use combustible fluids to start your fire. Instead, build one like this: Place two logs horizontally on the stove bottom, grate, or andirons and lay tinder (shavings, tiny twigs, paper) in between. Place a handful of dry twigs of split kindling over the tinder, cover with small, dry logs, then build a "teepee" of kindling on top to sustain the fire until the logs are glowing. Remember to clean your grate regularly. And, take care that hot ashes left after a fire can't start another—unwanted—blaze in the middle of the night. When fires are out, close the damper, or all the money you have saved will go up the chimney.

Stovepipes and Chimneys

Installing a stove   When installing a stove or Franklin fireplace, keep the chimney and the stovepipe connecting a stove and chimney short with as **few bends** as possible. Three 90-degree bends should be maximum. No bends is best.

Pipe joints should be tight enough so you can't see light through them. There should be no cracks or holes in the pipe or chimney. Corroded pipe should be cleaned or replaced. Assemble the pipe so that higher sections fit into lower sections rather than vice versa. This lets liquid creosote drip back into the firebox where it can be burned.

The experts also advise to use only **one pipe section** for horizontal runs. Be sure it does not sag. And, where a stovepipe goes through a wall, floor or roof, it should pass through an insulated thimble and collar to prevent fires. Check with your local building code agencies for approved installation procedures.

Keep as much of the chimney as possible **inside the house.** You pick up extra heat this way and reduce the possibility of creosote condensation caused by a cold chimney. Be sure to install a chimney cap to keep rain and snow out.

If you have ever been through a chimney fire, you will agree that it is a frightening experience. Even if such a blaze does not seriously damage your house, it can scare you out of your wits, bring the fire engines roaring to your address and drive your pets to leave home—this time for good.

What starts most chimney fires? Wood improperly seasoned and dried will produce **soot and creosote** when burned. When this accumulates in the chimney flue, all it takes is a roaring fire in the fireplace or stove to ignite it in a tremendous chimney fire.

Chimney maintenance   How often should you **clean your fireplace chimney?** If you only build a fire occasionally, the flue may not need cleaning for many years. However, if you use your fireplace frequently, especially in very cold weather, or if you burn softwoods or green wood, which produce a lot of creosote and soot, rather than well-dried hardwoods, (which produce little of either,) your flue may need cleaning once a year, or more.

You can easily find out what shape your chimney is in by **inspecting it** with a flashlight from the fireplace or the roof. The inside won't be spotless, but if it is covered with a thick, furry coat that flakes or crumbles when you brush it with a stick, you know the time has come for a thorough **chimney cleaning.** The simplest way to clean a flue is to hire a chimney sweep for $50 or so per flue to do the job. Look in your Yellow Pages under Chimney Cleaning.

However, if you don't mind climbing the roof and getting quite dirty, you can do the job yourself. The object is to loosen the soot and creosote so that it all falls into the fireplace. Here it is scooped up with a shovel or dustpan, or pushed down into the ash pit. First, you will open the fireplace damper, then close the glass doors or cover the fireplace opening with heavy plastic so the sooty mess doesn't get all over the house.

There are several easy ways to loosen soot and creosote. One of the most acceptable is to roll a short piece of **chicken wire** so it is slightly larger than the inside of the chimney. Staple the roll to a long pole and force the chicken wire down the chimney. Pump the pole up and down a few times and the roll of wire will act like a brush on the inside of your chimney.

Still another way to clean a chimney effectively is suggested. If the masonry chimney is long and straight, fill a heavy **cloth bag** with straw and weight it down with a rock. Tie a rope to the end of the bag and lower it down the chimney to scrape the creosote off the sides.

You can also use a small **evergreen tree** to clean the chimney. Tie a weight to the top of the tree and a rope to the end of the trunk. Lower the tree down the chimney. As you pull the tree up, the branches will brush the chimney clean.

If your chimney has a cap which prevents most methods of cleaning, a **chain** might be the only possible way to do it. In this case, tie a rope to one end of a short chain—an old tire chain is excellent—and lower it down the chimney. If you can get the chain swinging, it will scrape loose matter off the chimney flue. However, be careful. If you swing the chain too hard, it may crack the flue tile or otherwise damage the inside of the chimney.

Whichever method you use, be sure the rope is very strong and tied securely. If it breaks, you might spend hours trying to remove your cleaning tool—the chicken wire or evergreen tree—which has become lodged in the chimney.

CHAPTER 7

Homeowner's Guide to Solar Heat

FREE FUEL from the sun sounds like a delightful idea, especially when fall and winter bring higher heating and hot water bills. Over two hundred companies now manufacture solar heating and solar hot water equipment for home use.

Arguments for Solar Heat

Barry Commoner, in his recent series for the *New Yorker* magazine, contends that "the sun will eventually be the dominant source of power because it is the only energy alternative that is all of these—renewable, environmentally benign and potentially economical."

Where will the solar-heating boom be most effective? It seems to have a good start in the Midwest, perhaps because that area can't live without fuel to heat. Whether or not solar heat can work for you depends on more than the number of miles you live from the equator. Ironically, Miami, for example, gets only 2,800 hours of sunlight per year; and Fargo, North Dakota, gets the same number of hours of sunlight. The maximum sunlight hours is considered to be 4,400. (Chicago gets about 2,611 hours of sunlight, which would be considered good in solar-heating terms—not the very best, but not poor, either. However, Chicago homes need about seven times more heat than those in Miami, although the people in Florida need considerable energy to keep cool, which is another facet of solar energy being exploited.)

Some projects have been completed that have demonstrated

we can get all the home-heating energy we need from the sun, even in northern climates. For instance, a home built by the Canadian Government in Saskatchewan as a demonstration of solar potential is 100-percent heated by the sun. This was made possible by the design of heat-storage facilities to retain the sun's heat. And, of course, good insulation is a "must" for every kind of efficient heating method, especially solar.

Larry Dieckman, a partner in The Hawkweed Group, a Chicago-based firm which builds solar homes, says, "A house's solar aspects take some getting used to. Solar homes don't end up looking like the homes people are used to." (His firm takes its name from the hawkweed, a distant relative of the sunflower that looks like the sun, and that "tracks" the sun in its daily route through the sky, always facing it.)

In Chicago, Hanson Park School is a solar-heated elementary school. The mounting of solar collection panels allow the school to be heated by solar energy 30 to 60 percent of the time. A conventional gas-fired furnace is used as a backup. The solar installation added $350,000 to the construction cost, but it is estimated to be recouped over a five-year period.

Solar Greenhouses

The **attached greenhouse** is becoming a popular method of getting solar heat into your home. With this system, and atrium, greenhouse or sun room is attached to the house on the south side to collect and store solar heat and share it with the rest of the house. Floors and a portion of the walls in an attached greenhouse are usually made of heat-absorbing masonry to store the heat. Heat distribution is easily attained by opening doors and windows into the sun-space, or by adding vents to upstairs bedrooms to allow the solar-heated air to rise into them. Many people grow their own flowers and vegetables in such a greenhouse. However, don't forget the doors and windows to control this isolated sun-space. Some really sunny days might get too hot in this glassed-in space, and others may not be warm enough to keep your plants healthy.

What Will Solar Systems Achieve?

"In Chicago, a good solar system should provide 60 to 70 percent of [a home's] total annual heat load. A conventional furnace must provide the rest," says Solar Consultant Bill Behles. For a fee of $25, Behles will analyze a building's roof slope and southern exposure, and give his client a complete solar-energy survey of the building's thermal efficiency. He estimates the building's yearly and monthly heat-load, the heat required to maintain a desired interior temperature. On the basis of his survey, Behles then proposes various possible solar systems, and provides cost and saving breakdowns. If a client decides to go ahead with a specific system, Behles sends a trained crew to make the installation. For **solar consultants** in your area, look in the yellow pages of your telephone book.

Before investing in a solar system, there are several factors you should consider.

• Does your local **building code** permit you to alter the shape of your roof with solar collectors?
• Does your roof get **enough sun** to make such a system work for you? And, is your solar collector likely to be blocked in future years by your neighbor's trees or a high-rise apartment building?
• Will the addition of a solar system increase the **tax assessment** on your home?

The Solar Energy Industries Association predicts that nearly four million solar heating units will be installed annually by the early 1990s. As it is, the association estimates that 210,000 such units were installed in homes or businesses during 1979.

The Carter administration has projected that solar use will more than triple in the United States by the year 2000 as part of its proposed energy-independence program. The administration wants to increase solar use from 6 percent to 20 percent of total energy used by the end of the century.

Insulation for Solar Heating

"Good insulation is vital to solar heating," says W. A. Shurcliff, physicist and solar expert in Cambridge, Massachusetts.

The estimated **payback time** for complete insulation suitable for solar-heated houses is approximately five years. Paul Neuffer, who is building one of the country's first all-solar subdivisions near Reno, Nevada, favors Styrofoam brand insulation. He says, "It eliminates the thermal weak spots that occur with most other insulation systems, reduces heat loss through the walls and foundations, and helps cut down on air infiltration."

Kinds of Solar Heat Systems

According to Sandy Kraemer, a Colorado lawyer who specializes in legal problems pertaining to solar energy development, says in his book, *Solar Law* (1978: McGraw-Hill): "An active solar system has three components—collectors, storage and distribution. A major factor in an efficient active solar heating system is how well it encourages the sun's energy to enter the system, and how effective it is at preventing escape of energy."

Active vs. passive systems   What is the difference between **active** and **passive** solar heating? Active solar heating requires construction of specially-designed solar panels to catch the sun's rays. Passive solar heating uses the floor, the walls, the whole house as a sort of big, lived-in trap for the energy of the sun.

With some solar-heating systems, shallow, box-like collectors fitted with pipes are mounted on your roof. Water or air runs through the pipes, absorbing the heat from the sun and transferring it to a storage unit, generally housed in the basement. Here the heat is contained in a large tank filled with water or rocks, which holds the heat ready to be circulatedback through the house. Enough heat can be stored in this way to last from two to five cloudy days, depending on the size of the system, the season of the year, etc.

Passive homes favor a building concept that works naturally, without need—in many cases—for the usual heating/cooling gear. And, unlike active systems, passive houses do not need expensive heat exchangers, pumps, etc. Passive technology is here to stay, and many homes using passive solar heat can provide 30 to 100 percent of heating needs, records show, even in our coolest states.

Types of passive systems   Passive solar homes have four basic requirements: solar collection, storage, distribution and control.

The simplest of all passive solar homes are heated through **"direct gain."** Here the home collects sunshine through large windows facing the south to warm the living space directly. The direct-gain home then stores solar heat in thick, massive floors or walls (they might be stone, concrete, brick, adobe, or even water containers) to hold the heat for use on cloudy days and at night. The solar heat is distributed throughout the house, mainly by radiation from the warm floors and walls and by convection as warm air rises into other spaces. Two controls are important in solar houses heated by direct gain, one against too much heat-loss and the other against too much heat-gain in summer. The control to prevent excessive heat-loss can be in the form of insulated shutters, sliding panels or insulated draperies. The control to prevent excessive heat-gain in summer can be as simple as a roof overhang or leafy trees.

"Indirect gain" is another type of passive solar heating. With this method, the sun's rays do not have to travel through the living spaces to reach the heat-storage mass. And, the homeowner no longer sees the equipment for the collection of heat, only for its storage and distribution. This system also allows the house to collect heat at much higher temperatures than the direct-gain method does, without possible chance of overheating.

Kinds of indirect gain systems   There are three popular types of indirect-gain passive solar systems. The first type is the **mass trombe,** (named for its developer, a Frenchman named Dr. Felix Trombe.) Here, the sun's rays are absorbed directly behind a large south-facing collector area—including a lot of glass—by a massive wall (oftentimes twelve inches of concrete), which serves as the solar storage bin. This stored solar heat is distributed by radiation from the inside of the wall eight to twelve hours later, depending upon the thickness of your storage wall. Many times this trombe wall will be interrupted by windows and vents at the top and bottom so that hot air between the glass and the trombe wall can flow into the house immediately for distribution by convection. These vents can also be used in summer to channel excess heat to the outside.

A second type of indirect-gain passive solar home employs water—in barrels, bottles or bags—for solar storage behind the south-facing glass wall or collector. In this **water-wall system,** you will find solar heat distribution by radiation is much faster than in the mass trombe house, since hot water circulates very quickly, making the inside of the wall immediately warm and radiant.

The third indirect-gain passive solar system has **water storage reservoirs** on the roof. However, this method is a bit difficult to control. There must be some means of protection for heat-loss on a cold winter night and heat-gain on a bright summer day. For this reason, some type of hinged insulating panels or automatic insulating doors must be added. At the end of a sunny winter day, the cover is closed to keep the heat within the house, and on a cool summer night the cover is opened in order to chill the water down and give you a nice cold pack atop your house for cooling the next day.

Requirements for passive solar systems   Several factors are necessary to make the heat trap in passive solar heating work:

1. **Good insulation** and a tight, draft-free house. (This is necessary for any type of heating system to work effectively.)
2. **Southern exposure** of the major window installations.

Even though the sun might be low in the sky during the winter, quite a bit of the sun's light will enter a home with good window exposure to provide plenty of heat.

3. Some kind of **"thermal mass"** is needed to absorb the heat once it gets inside the home. Stone walls and floors should be dark-colored, since the darker they are the better they will absorb light and heat.

4. Some **device to trap the heat** gained through the windows after the temperatures drops at night, such as insulated draperies or shutters.

5. An **overhang** outside the house to shade windows during the summer to keep the home cooler. When the sun is higher in the sky during summer months, this overhang will prevent it from shining in the windows and heating up the house.

6. **Solar windows** facing south and east, should be made of double thermal pane glass. All skylights, which also admit sun and heat during the day, should be equipped with insulated sky-lids which automatically open and close to prevent nighttime heat-loss.

7. If glass walls allow in the sun's heat by day, they should be insulated by night with **roll-up shades** consisting of five layers of Mylar. These shades inflate and expand when warm air strikes them.

8. Also essential to a passive solar plan is a twelve-inch-thick **brick wall** which is the spine of the house. Rooms are "plugged" into this central "heater" wall, which stores excess heat from the sun.

9. Cylindrical **water containers** will also be positioned against a glass wall, to store heat during the day when the sun hits them, and release it at night. In a two-story house or bi-level, pipes in the stairwells will take heat from the ceiling and return it to the storage wall.

As you can see, the passive solar home is an actual heat trap. Passive solar heating can't achieve the same high temperatures an active system can, but that amount of heat isn't usually needed in a passive home. Passive homes tend to radiate heat throughout the home from one warm structural surface to another, and radiant heat feels warmer than forced-air heat does at identical temperatures.

Passive Solar Cooling

There are many passive cooling techniques being used today, and many of them have been around for quite some time. Perhaps you have used **evaporative cooling**—using water in hot, dry climates to pull heat from the air, which not only reduces its temperature, but increases humidity. Then there is **dessicant cooling** (using salts to pull humidity from the air, making the air much more comfortable in hot, humid climates). And **"time-lag cooling"** has been around since the Indians first invented it. This is the use of heavy adobe or concrete walls to delay the daytime heat's traveling into your home.

Passive solar housing is even being proclaimed a success in regions like Santa Cruz, California, where nighttime heating is needed year-round because of the cool ocean climate. Winters are cloudy and rainy, and fog is common all year. Studies show that even without full sun for long stretches of time, the passive solar system can provide 98 percent of the heat needed in winter, and 100 percent of the cooling needed in summer. And, best of all, this solar heating and cooling doesn't put a penny on your utility bill!

Solar Water-Heating Systems

Many new homes buyers today have the option of a solar water-heating system. (These systems have been used successfully in homes in Northern Illinois for more than four years.)

Solar panel system   One such system is called **Solarcraft IV** and is manufactured by State Industries of Tennessee. It has three to five 30-by-90-inch solar collector panels installed on the roof, and a 120-gallon water storage tank in the home's basement or utility area. However, many similar systems are available (check your area). The cost of installing the system in a home is about $4,500, but homeowners can cut the cost with tax credits provided for by the National Energy Act. Taxpayers can receive a credit of 30 percent of the first $2,000, and 20 percent of the next $8,000. At that rate, a credit of $1,100 would be allowed on a $4,500 expenditure. And, that's a tax credit, not a deduction. It's much more valuable than a deduction, since the whole amount of the credit comes off the bottom line of your taxes due.

How do these solar water-heating systems work? As energy from the sun is absorbed by the solar collector panels, distilled (and as a result, less corrosive) water from the storage tank is circulated through the collector panels, where it picks up the heat. The distilled water is then pumped through a piping system to sealed chambers within the storage tank. The heat is transferred from the chambers to the family's usable water surrounding them. When the heat from the panels has been transferred from the solar-heated distilled water to the storage water, a pump recirculates the distilled water to the collector panels for more solar heating. The process continues as long as the sun is providing heat and the thermostat in the storage tank is set for more hot water. When the sun is not providing heat, a standby electric booster in the storage tank heats the storage water to the desired temperature.

In addition to the tax savings, solar water heating can save as much as 60 percent of the conventional fuel needed each year to heat water for a family of four using the three-panel system in the Midwest. This calculation, which was computed at the University of Wisconsin's Solar Energy Laboratory, using actual weather data and electricity prices for the area, is based on an estimated average consumption for such a family of 118 gallons of hot water per day.

Solar concentrator system   There's still another concentrator for hot water heat. Called a **solar concentrator** because it multiplies the sun's energy by gathering a wide angle of light rays, the collector can be fixed to a roof to trap energy for heating and cooling or making steam for industry. But at this point it is found extremely efficient for heating water in our homes. Sunmaster Corporation of Corning, New York has begun manufacture of this device under a licensing arrangement with the U.S. Energy Department, and the design has been under development at Argonne National Laboratories near Chicago since 1974.

Perched about twelve feet off the ground on a metal platform that resembles a bandstand, the solar concentrator consists of two modules, each about the size of a coffee table top. Viewed from the side, each module consists of parallel channels, or troughs, lying side by side. Glass tubes in the center of each trough conduct water in and out of the concentrator. The inside surface of each trough, lined with a material similar to aluminum foil, reflects light rays toward the tubing, thereby heating the water. And William Schertz, Argonne Laboratories' program manager for solar applications, says, "It is the shape and reflective surface of the troughs that account for the concentrator's efficiency in diffuse light—temperatures up to 350 degrees can be achieved even in climates where gray and hazy days dominate."

The cost of Sunmaster's concentrator is approximately $22.50 per square foot of collecting area. Thus, for a family of four in Illinois, approximately 56 square feet of collector—for a cost of about $1,260—would mean the solar system would meet between 60 and 80 percent of annual hot water needs, Schertz claims.

Underground Houses

Underground Passive Solar Houses are also coming into their own. Some of these are known as "cave houses." As solar power catches on with the public, Dennis Blair, an architect who trained with Frank Lloyd Wright, says, "Some of these homes will be built into hillsides with skylights and lots of south-facing windows to compensate for the lack of light coming into the house from other directions."

"Berming", or piling earth up around the outer walls, cuts

the amount of space in the house exposed to cold winter winds, which in conventional houses is a major factor in heat-loss. In addition, the "underground" or "cave" house would keep the sun's heat in winter and keep it out in summer.

Charles Lane, assistant to the director of the Underground Space Center at the University of Minnesota, estimates that today as many as two thousand underground homes have been built or planned. In addition, schools, libraries, offices, and other structures also have been built into the ground.

For instance, in Chicago a $5.2 million addition to Westinghouse High School aimed at a 35-percent reduction of energy consumption. The addition houses a double gym, Olympic-sized swimming pool and theater. It was sunk 17 feet below the ground for insulation, and has no windows.

According to Malcolm Wells, a Massachusetts architect who has made monumental strides in creating interest in underground building, "Underground buildings simply do not have to fight the large temperature fluctuations above-ground buildings encounter. Many people envision underground homes as dark, slimy places filled with spiders and rats," Wells adds. "While some buildings are like vaults, most have windows or skylights to allow the buildings to fill with sunlight."

As in all construction, **insulation** is a "must" for the underground home. For instance, the Dow Chemical Company has found that below-ground heat-loss through foundation walls can amount to as much as 20 percent of total heat-loss in a typically insulated two-story conventional home, and 25 percent loss in a one-story conventional atop-ground home. In below-ground houses the potential for such heat-loss is, obviously, even greater. Styrofoam brand insulation is the most widely used for insulation in underground construction.

The earth's temperature at the eight-foot-below-ground level is a constant 58 degrees, with only a slight variance in some areas. If it were 10 degrees below zero outside, you would need to heat a conventional home to 80 degrees to reach 70 degrees inside. In an underground building you would only need to increase the temperature 12 degrees to reach the same temperature. Of course, keeping cool in the summer is no problem in an underground house. Underground houses are apt to be coolest in May, for it takes three months for the earth to respond to temperature changes brought on by seasonal variations in heat and cold.

Not only do underground buildings save money on fuel, they are virtually fireproof because they are made of concrete. Many are eligible for discounts on insurance.

What about cost? Wells says the cost of building underground is about the same as building above ground, depending on the site. "Conventional materials are used in conventional ways in building underground structures," he says.

Tax Breaks and Costs of Solar Systems

To stimulate solar heating, at least thirteen states—Arizona, Colorado, Illinois, Indiana, Maryland, Massachusetts, Montana, New Hampshire, New Mexico, North Dakota, South Dakota, Oregon, and Texas—have enacted legislation giving homeowners tax breaks if they install solar equipment. And, the number of participating states is growing monthly. Check your state for latest details.

Studies show solar heat is **competitive with electricity** in areas where electricity costs more than four cents per kilowatt-hour. George O. G. Lof, pioneer solar-energy authority and director of the Solar Applications Laboratory at Colorado State University says, "If your gas or oil heating costs are in the $200 to $300 annual range, there would be no incentive for using solar heat. However, with many people's yearly heating costs now rising above $500, and, in electrically-heated homes in some localities, even above $2,000, a savings of up to 75 percent by use of solar heating is definitely attractive. In the next fifteen years, as conventional energy becomes more scarce and more costly, I expect we will have hundreds of thousands of solar-heated buildings." However, cost of installation is an important factor.

Things to Remember about Solar Heating-Cooling Systems

• **How long** should your solar system last? The experts claim your collector and storage unit should last for the life of the building. If it doesn't, the design isn't what it should be.

• Going solar doesn't mean you have to **change** the way you live or go out and buy lots of mod-looking furniture. Inside, a solar house looks pretty much like any other house, except that windows are built to flood the rooms with the maximum amount of sun.

• Although no power company can meter the sun, solar energy is **not exactly free.** If you design a new house to take advantage of passive solar energy, for instance, your dollars will go for extra insulation, double layers of glass and special design features.

• The payback period is just one factor to consider in deciding whether to go solar. The first consideration, obviously, is whether your house gets **enough sun** to power such a system.

• Solar energy can provide us with over **one-fourth of all energy needs** by the year 2020, say the experts, safely, cheaply, and without polluting. It's going to be the coming thing, since use of solar energy systems could release the finite fossil fuels—coal, gas, and oil—for areas where there are no alternatives to their use.

• We aren't the only country going to solar energy. **Japan** has more than a million homes making use of solar hot water heaters.

Before you buy solar equipment   Solar heating and cooling is still a virtually new field. Consumer protection is still minimal for buyers of solar heating systems. Here are some guidelines provided by the Office of Consumer Affairs (published by the Federal Energy Administration) in a book, *Buying Solar:*

1. Ask for **proof** that the product will perform as advertised—a report, say, from an independent, reputable testing laboratory or university. Ask your engineering consultant to go over this report.

2. Examine the **warranty** carefully. What are the limitations? How long does the warranty or guaranty last? Are parts, labor and service covered? Who will provide the service? (Don't settle for a promise that any plumber or handyman will do.)
3. Be especially careful or leery when buying a solar heating system piecemeal. **Solar components** are like stereo components—some work well together, others do not. Don't try using a do-it-yourself kit unless you have a solid mechanical background.
4. Ask the seller for a list of **previous purchasers** in your area, then consult them. And check on the seller with your local consumer-protection agency or local Better Business Bureau regardless of the reputation of the seller.
5. Beware of **fly-by-night sellers** who use post-office box numbers. Find out from the seller how long he has been in business, where, and what his financial references are.

As with any major investment, solar heating and cooling deserves the most careful study and investigation before you buy.

Solar energy presents many problems yet to be solved. There are questions of engineering and economics, and even of legal "sun rights" (though someone has yet to criticize sunshine as a source of pollution). The sun is the only perfect energy source, solar advocates say. Some experts also see solar energy as the best long-term alternative to nuclear power, with coal as the answer in the meantime.

When you are ready to talk to an experienced designer for an estimate on making your home solar heated and cooled, call the **National Solar Heating and Cooling Information Center,** using their toll-free numbers (800-462-4983 in Pennsylvania; 800-523-4700 in Hawaii and Alaska; and 800-523-2929 in the remaining states) and ask for a list of current practitioners and the addresses of homes they have built in your part of the country. Talk with the designers and be sure to visit at least one of their homes.

As you talk with the designer, remember he will be working with some major factors—your climate, temperature, wind, humidity and sunshine factors.

Factory-built solar heating systems may now be marketed

in your area. Look in the Yellow Pages under Solar Energy Equipment.

Find out more about solar energy by writing

The American Institute of Architects, 1735 New York Avenue, N.W., Washington, D.C., 20006, or by requesting *Buying Solar,* c/o Joseph Dawson, Office of Consumer Affairs, HEW, Washington, D.C. 20201.

The National Solar Heating and Cooling Information Center, P.O. Box 1607, Rockville, Maryland 20850.

Consumer Action Now, Inc., 49 E. 53d Street, New York, New York 10022. Ask for information and lists of solar architects, engineers and manufacturers.

Superintendent of Documents, Government Printing Office, Washington, D.C. 20402. Ask for *Buying Solar,* a wonderful guide to choosing solar systems, with bibliography, tables of weather data and cost factors; and ask for the *Catalog of Solar Energy Heating and Cooling Products* (ERDA-75) which carries a list of manufacturers— (Document #0520010-00470-1, $3.80).

Environmental Information Center of Florida, Conservation Foundation, 935 Orange Avenue, Winter Park, Florida 32789. Ask for *"Build Your Own Solar Water Heater."* This is for do-it-yourselfers and anyone trying to evaluate manufactured solar water heaters ($.50).

Solar Energy Industries Association, Inc., 1001 Connecticut Avenue, N.W., Washington, D.C. 20036. Ask for information on guide to solar products.

Edmund Scientific Company, Edscorp Building, Barrington, N.J. 08007. Ask for material on solar devices for do-it-yourselfers and complete plans for solar houses.

CHAPTER 8

Saving Energy with a Heat Pump

THE HEAT PUMP is not new. The first factory-made heat pumps rolled off production lines more than twenty-five years ago. More than one million heat pumps are already in operation in nearly every section of the country—in offices, schools, motels, stores and industrial plants as well as homes. Despite its name, a heat pump is designed to provide summer cooling as well as winter heating. And they are gaining new popularity due to extreme focus on energy conservation.

Is there a heat pump in your future? According to the Air-Conditioning and Refrigeration Institute (ACRI), "There could very well be, for this impressive device not only will heat and cool your house for less, it promises major savings on monthly energy bills for many homeowners, plus hope for easing the depletion of our nation's critical fuels."

What is a Heat Pump?

It is basically a central-air-conditioning system that can both **heat and cool.** In the warm months it acts like a standard air-conditioner to remove hot air from your home. In the colder months, the heat pump extracts heat from the outside (even on the coldest of days there is a good deal of heat in the outside air), raises its temperature, and then pumps the warm air into your house.

The heat pump is gaining in popularity because it uses **free heat outside,** and can heat a house much more efficiently than a conventional electric furnace. A conventional furnace may be used for supplementary heat on very cold days, when there may not be enough heat in the outside air for the pump to function efficiently.

How Does a Heat Pump Save Energy?

"It's the **heating cycle** that accounts for the significant energy savings that are produced by heat pumps. Unlike a furnace that turns fuel or electricity into heat, the heat pump collects heat that already exists in the outside air by means of its refrigeration cycle. This means that the heat pump can supply from one-and-a-half to two-and-a-half times more heat than the energy it uses," says the ACRI. The heat pump can do this because heat exists in air even when the winter cold is down to minus 460 degrees F.

"For example," the ACRI continues, "a heat pump can mean savings of 30 to 50 percent on electric heating bills because it uses 30 to 50 percent less energy to supply the same heat as an electric furnace or a resistance heater. Engineers refer to this advantage of the heat pump as the efficiency or **Seasonal Performance Factor** (SPF). The higher the SPF, the more efficient the unit."

If you live where wood, gas or electricity are precious (and that seems to be almost everywhere these days), **supplement your forced-air system** with a heat pump, which runs both hot and cool. In a nutshell, the pump pulls warmth from the outside air in winter and from the inside air in summer. The large appliance company that makes it says the pump can often handle the entire heating load by itself—your furnace steps in only when the temperature really plummets.

A heat pump is an especially good choice if you are **replacing** both the heating and central cooling system. But, even if you just want to add or replace a central air-conditioning system, a heat pump may be a wise choice.

Remember, a heat pump can be installed in an **existing house** that already has duct work for a forced-air heating system. The amount it can save you will depend upon such factors as climate and local energy costs. Your utility company can help you estimate what it would cost to operate a heat pump in your particular area.

During the heating season, the heat pump's Coefficient of Performance (COP) increases on mild days and decreases on cold days (see the heat pump glossary at the end of this chapter). The average COP for the heating season is, therefore, higher in a mild climate than in a region where winters are severe. For this reason, many of the early heat pumps were installed only in southern sections of the country.

Through the years, improvements in design have broadened the geographical range of heat pumps to almost *every* section of the country. While the seasonal performance factor of a heat pump will be lower in a cold-winter area, the heat pump will still be more efficient than other electric heating systems in that area, due to recent steep rises in utility charges.

Models and Sizes

There are many different heat pump models and sizes to select from. The first thing you need is a knowledgeable dealer who will accurately estimate the correct-size heat pump for the heating and cooling requirements of your home. One manufacturer has a special computer system to help the dealer determine this. For maximum energy savings, the heat pump also has to be installed and maintained properly. And like any heating or cooling system, a heat pump performs best when used in a well-insulated house which is equipped with storm windows and weather stripping.

Most heat pumps are compact units that, except for indoor components, are **installed outside the home.** In size and appearance they look like the outdoor unit of a central air-conditioner.

In a typical heat pump installation in a home, the outdoor unit contains the outside coil, compressor and reversing valve. The refrigerant travels through pipe or tubing to the inside coil located in the path of air circulated by the inside fan. The supplemental electric heater above the inside coil is activated when the heat-loss of the building exceeds the heat pump output on coldest days.

Costs

On the average, heat pumps have a higher initial cost than other heating-cooling systems (about $2000 for a whole-house unit, down to $425 for room size). Why? "This higher cost is a reflection of the durability that must be built into the heat pump for all-year-around operation in hot and cold weather, and of the heat pumps's sophisticated control mechanisms," says the ACRI.

A recent Federal Energy Administration (FEA) study analyzed the cost of installing and operating a heat pump in nine cities. Costs were based on actual contractors' estimates and local energy costs for a heat pump installed in a two-story frame house with 1,850 square feet of living space, occupied by two adults and two children. (The hypothetical house had storm windows, and insulation in attic and outside walls that met with the Federal Housing Authority–Department of Housing and Urban Development Minimum Property Standard for each location.)

In the nine cities studied, the FEA found that yearly energy costs for a heat pump system were consistently **less** than for an electric furnace-central cooling system. However, we must remember that actual energy costs vary *widely* by region. Your local utility can help you estimate what it would cost you to operate a heat pump in your area.

Living with a Heat Pump

You will find your heat pump runs longer than the typical furnace, because the heat pump delivers air at temperatures closer to room temperatures than fossil-fueled heat systems, even though both systems deliver the same total amount of heat to your home. You will enjoy the fact that the heat pump produces **less temperature fluctuation** in your home during the day. And, since the system is flameless, the heat pump produces **no fumes, soot, or smoke** to soil your home furnishings and walls.

With the help of the ACRI, here is a list of things to consider when buying a heat pump:

"**Initial cost**   Get two or three contractors to estimate the installation cost of a heat pump for your home versus the cost of an alternative heating-cooling system for your area.

"**Number of years to recoup your initial investment**   Figure out how many years it will take for your heat pump to pay back the higher initial cost with lower annual operating costs. You can do this yourself by dividing the estimated annual operating savings into the extra cost you would pay for heat pump installation.

"**Service and maintenance costs**   The availability of expert service and maintenance is just as important with a heat pump installation as it is with any other heating-cooling equipment. Make sure the contractor you deal with is an authorized heat pump dealer. Check to be sure his servicemen have attended manufacturers' or industrial training schools on installing and maintaining heat pumps. Also, ask about an extended warranty program for your heat pump.

"If you decide the energy savings will offset the initial cost of a heat pump in a reasonable length of time, and that your contractor will perform reliable and professional service, then the heat pump will probably be for you."

Heat Pump Glossary

When purchasing or shopping for a heat pump you will probably encounter much terminology which will be puzzling to you. Here is a tentative glossary applicable to the purchasing of a heat pump:

Authorized dealer A dealer who is authorized by the manufacturer to install heat pumps.

Air-source heat pump This is a heat pump that transfers heat from outdoor to indoor air during the heating cycle—the most common type of heat pump installed in homes today.

Water-source heat pump This heat pump transfers heat from a water source to an indoor air circulation system. This is frequently installed in commercial heating systems.

British Thermal Unit (BTU) The amount of heat required to raise the temperature of one pound of water by one degree Fahrenheit.

Balance point This is an outdoor temperature—usually between 20 and 45 degrees—at which the heat pump's output exactly equals the heating needs of the house. Below this balance point, supplementary heat is needed to maintain indoor comfort.

Coefficient of Performance (COP) A ratio calculated by dividing the total heating capacity provided by the heat pump's refrigeration system—including circulating fan heat but excluding supplementary resistance heat—by the total of electrical input (watts) and multiplying the result by 3.412. (A ratio calculated for both cooling and heating capacities by dividing capacity in watts by power input in watts.)

Energy Efficiency Ratio (EER) A ratio calculated by dividing the cooling capacity input by the power input in watts at any given set of rating conditions, expressed in BTU/hour per watt.

Heat sink A body of air or liquid to which heat removed from the home is transferred. In a heat pump, the air outside the house is used as a heat sink during the cooling cycle.

Heat source A body of air or liquid from which heat is collected. In a heat pump, the air outside the house is used as a heat source during the heating cycle.

Heat transfer The process of transferring heat from one location to another.

Inside coil The portion of a heat pump that is located in the house and functions as the heat transfer point for warming or cooling indoor air.

Outside coil A portion of the heat pump, located outside the home, that functions as a heat transfer point for collecting heat from or dispelling heat to the outside air.

Seasonal Performance Factor (SPF) The measure of the efficiency of heating equipment over the length of the heating season.

Split system heat pump A heat pump with components located both inside and outside of a building—the most common type of heat pump installed in a home.

Supplementary heat At temperatures below the heat pump balance point auxiliary heat must be provided. In most cases this is done with electric elements which are a part of the heat pump system installation. A gas or oil furnace could also be used to provide supplementary heat.

Watt A measure of the work energy of electricity.

CHAPTER **9**

Humidifiers and Dehumidifiers for Your Home

THE MOST OBVIOUS way to humidify the household air is a way we have all tried—creating steam. You will notice that the bathrooms and kitchen seem to be the most comfortable parts of the house in winter because that's where most of the steam is created. Some people put shallow pans of water on radiators or on registers, but rarely do these areas get hot enough to give off much more vapor than the water would have given off with normal condensation.

Humidifiers

To stay comfortable at a lower temperature, you might want to use a humidifier, either on the furnace or a portable model. Winter is the season when most homes can benefit from a humidifier. Cold air can't hold much moisture. When heated, it can hold more, and proceeds to absorb moisture from the surroundings. A humidifier restores enough moisture to the air to bring the relative humidity to a comfortable level.

Dry air often causes higher heating bills, too, as the moisture evaporates from your skin and makes you feel cooler and in need of more heat—so, up goes the thermostat. With the humidifier, you will find you are just as comfortable at 68 degrees as you were at 73 degrees without it. Since, as we have seen, you knock about 3 percent off your heating bill for every degree you lower the temperature, the humidifier can mean a real savings in dollars.

Granted, it costs **energy** to add moisture to the air, but with windows closed, the air in your house is recirculating. And, once a specified level of humidity is reached, the humidifier turns itself off. Thus, once you can maintain a lower temperature because of a relatively high level of moisture, you will save on energy.

You will find excessively dry air will dry out your throat and nasal passages, making the sinuses painful as well as making your skin dry and flaky. It also creates static electricity to a degree that can be annoying when you touch a doorknob or another person. Air lacking in sufficient moisture is also hard on wood furniture and woodwork, drawing moisture from them and causing cracking and warping. It can cause furniture joints to become unglued, book bindings to crack, paintings to flake, and pianos to get out of tune. It's even hard on your pets and plants.

How Do Humidifiers Work?

Most types use similar methods of humidification. A porous filter or foam filter pad is wetted either by rotation through a water reservoir or by having water pumped or splashed onto it. Simultaneously, dry air is forced through the wet pad by a fan and picks up evaporating water. The moistured air flows out of the front or top of the unit and spreads throughout your room— or home.

In most console humidifiers, a **humidistat** automatically turns the unit off and on to maintain the degree of moisturization you select. Tabletop and room models without humidistats run continuously. There are also room humidifiers called **cool-mist atomizers,** that throw tiny water droplets into the room air. If the water is hard, you will find the minerals may deposit a fine white dust on your furnishings.

What Size Humidifier Should You Buy?

Be sure to buy the right size. Many consumers make the mistake of buying those attractively-priced table models, hoping they will be big enough to humidify a room. However, these small models, which sell in the $20 to $25 price range, seem to do very little, even if you are sitting right next to them.

As a rule of thumb, a floor console is designed to humidify several rooms, a tabletop model suits one or two rooms or a small apartment, and a bowl-shaped room unit has bedside or chairside capacity. Some inexpensive consoles have water-output capacities as low as five gallons per day, while some of the most expensive units can disperse up to seventeen gallons a day, and have many sophisticated controls and convenience features.

How Much Output Do You Need?

The average home requires roughly **one gallon** of water per day per room. However, a home in a severe climate, an older building, or a building lacking in storm windows or storm doors could need one-and-a-half times the average amount. On the other hand, a new, well-insulated house with double windows and doors may need only one-half gallon per day per room unless the rooms are unusually large and spacious. A large, active family, frequent meal preparation and showers, and home laundry, could reduce your humidification requirements appreciably.

For help in making a more precise estimate of your needs, you can get a free humidification selection guide from the **Association of Home Appliance Manufacturers** (AHAM), 20 North Wacker Drive, Chicago, Illinois 60606. (To use the guide effectively, you must determine the floor space in your home, and type of home construction (brick or frame), along with amount and type of insulation.

Before You Buy

Before buying, examine each humidifier you consider. Be sure you find a **safety symbol,** such as that of Underwriters' Laboratories, Inc., on the serial plate or on the unit itself.

If you find a separate seal saying "**AHAM Certified,**" the humidifier has been certified by the Association of Home Appliance Manufacturers, and you can compare that model's water output rating in gallong per day, listed on the serial plate, to the ratings of other AHAM-certified models. Some brands without the AHAM seal also list capacities, but the ratings may or may not be comparable.

Be sure to scrutinize the **warranties and service policies.** Most warranties extend for one year after purchase and offer free replacement or repair of defective parts. A few promise a new machine for an imperfect one. Whatever the warranty, be sure that **repair service** is available locally in your area so you can avoid the trouble and expense of shipping the unit to a distant factory or service center. You will find some dealers offer warranty privileges in addition to those of the manufacturer, and are also approved service centers. For the best bargain in any type, shop comparatively at several stores and watch for advertisements of sales.

Styling

In addition to being attractive, a cabinet humidifier should have **smooth corners** and edges with no cracks, crevices or warping between parts. If it is made of plastic, it should be sturdy in appearance. Grilles, louvers, and inside surfaces should be either plastic, aluminum, or metal specially coated to prevent rusting.

Water-level and operational **signals** should be easy to see. The outward flow of moist air may be efficiently directed by a top **grille** or adjustable louvers, a large grille in front, or concealed slots on three sides near the top. A console should rest on four free-swiveling **casters** for easy rolling from spot to spot.

Other Features

The **water reservoir** may be either part of the cabinet or a removable plastic tank. Since you must drain and clean the reservoir every few weeks, look for features that make the job easy. Hand holds on a removable tank should provide a firm grasp for handling. Rounded inside corners won't trap dirt. A drain on the built-in reservoir eliminates dipping out and sponging up water the humidifier can't utilize, and smooth inside edges and parts won't snag or scratch when you reach inside. Especially note if any parts must be removed for cleaning access. If so, do you need special tools to remove them?

Most models are filled by rolling them to a sink, so check the sturdiness of the **casters** and how well they roll. Some tanks are **easier to fill** than others. Ask the salesman to demonstrate how to fill each model you consider for purchase. Some are equipped with a tilt-out front, or water-guide panel (in some this is located at the back), to prevent spilling and splashing. Others contain a wide-mouthed, long-necked funnel. A few include a hose for attaching to a faucet. Remember, though, that the hose may eliminate splashes during filling, but won't prevent sloshing over the reservoir sides as the unit is returned to its working location.

You should also check the ease of cleaning and maintaining the **filter**. The filter eventually will become caked with minerals, dust and lint that will impede the air-flow. The general care recommendation is to clean the filter once a month and replace it once a year. Some filters are easy to remove, others require more effort. And, since any filter pad must be periodically removed for cleaning or replacement, examine each model to see how difficult or easy this task would be. If you have any doubts, ask the salesman to demonstrate the procedure. Ask where new pads may be purchased and if the original one may be washed and reused a few times.

What about the **motor?** Most humidifiers have just one motor to drive the fan and the other moving parts, but in some models a second motor drives a drum, or roller, or a water pump. Unless the motor or motors are permanently oiled, be sure the owner's manual includes clear oiling instructions for you to follow.

Since you should never use an extension cord to reach an outlet with your humidifier, the longer the unit's **power cord,** the greater the flexibility you will have in positioning a model in your home. (An extension connection is a fire hazard, as well as a shock hazard if water happens to be spilled on it.)

Optional Equipment

There are many convenience options available in humidifiers today. While these may be controls or built-in features that may increase a model's convenience of operation, they also raise its price. It is up to you to decide whether these options are worth purchasing:

Water-Level Gauge. Shows the amount of water left in the reservoir.

Fan control. Usually provides a choice of "high" fan speed for fast humidification or "low" for a more quiet operation. A "fan only" setting permits the air to circulate without moisturization.

Signal light. Glows when the reservoir needs filling. On some models there is still a second light that shines during operation.

Automatic valves. Control refill and overflow and allow the connection of a unit to household plumbing by means of hoses.

Automatic shut-off. Stops the unit when the water reservoir is almost empty and continued operation would waste electricity and cause unnecessary wear on parts.

Electric heater. Steps up water output capacity, but sharply increases the use of electrical current. As an energy-saver, this is *not* a good idea.

Removable power cord. Allows separate storage of the cord and the unit.

Summer cover. Fits over the top grille or louvers when the unit is not in use. Making your own cover, or improvising, would be much cheaper.

Maintenance

Read the **owner's manual** completely before filling or plugging in the unit.

When placing your unit, put it **near a room partition** rather than a cool outside wall on which water vapor may condense. Be sure to allow a good six inches behind the unit for air circulation. If your home has more than one level, the **bottom of the stairwell** is a favorable location. Avoid bathrooms or kitchens (they provide their own moisture), hot air registers, radiators, room corners, and areas which would be crowded by furniture.

For the first few days your humidifier is in operation, turn the humidistat and the fan control to a **high setting** to aid moisture absorption by furnishings as well as by room air. However, if the air feels too damp, or if you find frost or beads of water on the windows, select a lower humidistat setting. Too much moisture could damage the walls and woodwork in your home.

Once you feel the area is humidified, adjust the settings to your particular comfort level. Check periodically, especially during cold spells, for over-humidification.

Before **refilling, emptying, or cleaning** the water reservoir, turn the humidstat to the "off" position and also disconnect the power cord. This reduces the possibility of shock if you should spill the water. Be sure the area on the outside of the unit, your hands, and the cord are dry before you reconnect the cord.

Don't pour hot water in the reservoir, or overfill it.

During your usual household cleaning, dust the louvers

and grilles with a soft brush or dusting attachment of your vacuum cleaner. Dust the cabinet or wipe it with a damp cloth.

Every few weeks, empty the water reservoir and scrub the inside with a sponge or soft cloth and a mild detergent to discourage the growth of mildew, mold and bacteria, which could produce an odor.

Every month remove the **filter pad** and gently squeeze a mild solution of water and vinegar through it to remove mineral deposits and dust. After a full season of operation, check to see if you need a replacement filter pad.

Dehumidifiers

New, tightly-built, small homes tend to be too humid. Water may run down the windows, and the doors might stick if there is no proper moisture barrier in the walls. When this happens, remove excess moisture from the air. Besides causing discomfort, high humidity—like low humidity—also has damaging effects indoors. It can cause mildew on clothing, books and furniture; make doors or drawers swell and stick; cause tools and other metal objects to rust; make paint peel, cause walls and pipes to sweat; and produce an unpleasant musty smell.

Dehumidifiers pull the humidity out of the air, and are housed in neat cabinets. Water-removal **capacity** of various models ranges from eleven pints to more than thirty pints a day. There is no general way to advise what capability will be most effective, since climate, room size, and home construction all must be considered. A very rough rule of thumb, though, is that there should be at least one pint of water-removal capacity for each thousand cubic feet of space to be dehumidified.

Dehumidifiers, like humidifiers, also have an adjustable **automatic control** that you set at a desired humidity level. The appliance goes on automatically when the moisture content rises above the selected level, and operates until the humidity is lowered to that level. Many models shut off automatically when the water container is full, at which time a signal light goes on to tell you that the container needs emptying.

If you are not going to attach a drain hose to remove the water continuously, make sure the **water container** is easy to remove, carry, empty, and clean. Other features that are important to look for are the ease with which the applicance can be moved, its appearance, and the length of its power cord, since basements or storage areas often have less conveniently located electrical outlets.

CHAPTER 10

Insulation—The Energy-Saver That's Always in Season

ADEQUATE INSULATION and weatherstripping alone can cut annual heating costs by as much as 50 percent, the studies show! The basic purpose of insulation is to prevent heat-loss in winter and prevent heat from entering in summer months. It also helps to control moisture and, when properly installed, will effectively slow down the out-flow of moisture-laden household air and prevent condensation problems inside the walls.

Despite the savings that can be had with insulation, it is estimated that two-thirds of American homes are insufficiently or improperly insulated, and some have no insulation at all.

Insulation saves on fuel bills, according to some authorities, as follows: "Three inches of insulation in the ceilings of top-floor rooms can reduce fuel consumption by 34 percent; two inches under crawl-space floors can save you 15 percent; two inches in the walls will save another 10 percent. And, if you install four inches of ceiling and three inches of wall insulation, fuel needs can be slimmed by 47 percent; add an additional three inches in the ceiling, the most important place for insulation, and you have cut total fuel requirements to half those of an uninsulated house."

In the **attic,** you should have at least six inches of roll-type or nine inches of loose insulation to effectively reduce the downward heat-flow from the attic during summer and the upward heat-loss through the attic during the winter.

The average **new home** has insulation that meets the Federal Housing Authority's thermal insulation standards. (FHA standards are not legally binding, but they have a large influence on home building because of their widespread use for loan appraisals by conventional loan organizations as well as FHA.) If your house was built after June 1971, when FHA revised its insulation standards, it may be better insulated than those built before that date. Many homes fail to meet FHA standards.

John C. Moyers of the Oak Ridge National Laboratory, Oak Ridge, Tennessee, compared insulation required by FHA with insulation that yields the maximum economic benefit to the homeowner—that is, insulation that reduces the heating bill at least enough to cover the cost of the additional insulation and in most cases saves a good deal more. He found that with better insulation, a New York resident with a gas-heated home could reduce energy consumption by 49 percent a year, compared with insulation meeting FHA standards prior to June 1971. The owner of an electrically-heated home could reduce his energy consumption by 47 percent.

According to the Oak Ridge study, insulation that is the best bargain for the homeowner includes 6-inch ceiling insulation for electrically-heated homes, and 3½-inch ceiling insulation for gas-heated homes (revised FHA standards call for 3½-inch ceiling insulation for both); 3½-inch wall insulation (FHA standards call for 1-⅞-inch wall insulation); and floor insulation (also required by FHA.)

Vapor Barriers

One of the most important parts of any insulation job is the installation of a vapor barrier. Some blanket and batt types of insulation come with a barrier of especially-treated foil already attached. However, if you are using loose-fill insulation, or unfaced blankets or batts, you can make your own vapor barrier with sheets of polyethylene, metal foil, or asphalt sandwiched in brown paper. This should be installed toward the warm-in-winter side of walls, ceilings and floors. The vapor barrier **prevents moisture** in the household air from entering the insulation, condensing there, and destroying the insulation's value as a reducer of heat-loss. When you are installing the vapor barrier, carefully tape any tears in it so it will do the job for which it was designed.

R-Value

The "**thermal resistance value**" of any insulation material is measured in the terms of "R's." The higher the R, or thermal-resistance value, the greater the insulation will resist heat-flow through it. Take fiberglass, for example, which is recognized as an efficient insulation material. A fiberglass blanket just 3.5 inches thick (R-11) has the same thermal resistance value as a wooden wall 9 inches thick, a brick wall 4.5 feet thick, or a stone wall 11 feet thick.

In general, the higher the R rating of the insulation you buy, the more you will save. You will find insulation with the following ratings: R-7, R-11, R-19, R-30, and R-38. Minimum recommended attic insulation, for instance, is the R-19. Always select insulation by the R number printed on the bag—not by thickness, as thickness can vary by brand.

Types of Insulation

There are six basic types of insulating materials:

Blankets and batts   Made of fiberglass or rock wool with asphalt-paper or foil vapor barriers, these have the insulating material enclosed in a continuous envelope. Flanges on the sides of both permit them to be stapled to a frame.

Blankets and batts are similar in appearance but are sold in different forms. Blankets come in rolls of 40 to 100 feet in length and 15 to 23 inches wide to fit between standard 16- and 24-inch stud-joist-rafter spacings. These long rolls are cut to fit during installation. Batts are usually available in shorter lengths (four- or eight-foot lengths), by 15- or 23-inch widths for easy handling and for use in tight spots and cramped spaces. They usually range in thickness from 3.5 inches up to 12 inches.

Mineral wool (or rock wool), which is the most commonly-used insulating material, is available in blankets, batts and pouring wool.

Glass fiber insulating batts are also widely used. The glass fiber itself won't burn, but the asphalt-coated kraft paper often used as a backing can, if not treated to be fire-retardant. Although they cost more, some batts coming on the market now have the UL label, and this is an important advantage because they are fireproof.

A question frequently asked is, "Are glass fiber particles injurious to your health?" The National Mineral Wool Association asserts that while asbestos is a biologically active dust and thought to be associated with scarring of the lungs and malignant tumors, use of glass fiber has produced no evidence of chronic pulmonary disease, no fibrosis and no malignancies in man. The structural characteristics and chemical composition of glass fiber "are completely different from asbestos," adds the Association.

However, if you are installing glass fiber insulation, take some extra **precautionary measures.** For instance, wear a nose and mouth mask, as well as eye protection when you work with glass fiber overhead. This type of insulation causes skin irritation. After handling glass fiber material, wash with soap and warm water. Wash the clothing you wore for the job separately from other laundry.

Loose-fill insulation   Loose-fill consists of particles made of cellulose fibers, mineral wool, vermiculite, or perlite, which are poured or blown into place. Loose-fill is sold by the bag or bale, and it is especially suited for installation in existing houses where interior spaces may be otherwise inaccessible. For instance, it can be easily poured between the joists in an unfinished attic, then leveled with a rake for uniform coverage.

Since little processing and packaging is used in making loose-fill insulation, it is low in cost. However, you must remember that this type of material does not provide a **vapor barrier,** and it does settle over a period of years.

If you employ a contractor to install loose-fill costs range from 60 cents to 90 cents and up—per square foot of insulation installed, and savings could amount to 16 to 20 percent in utility costs.

Insulation board, Wood Fiber   Wood fiber board for insulating comes in sheets, panels, planks and tiles. It can be cut and fitted with regular tools and can be used by itself as an actual wall in new construction, or can be placed over old walls. With the combination of structural strength and insulating properties, it is especially good for ceilings.

Rigid insulation Boards Made from Lightweight Plastic Foam Foam boards are used for the insulation of basement walls and the perimeter insulation of floor slabs in new construction, and underneath siding on the house. Because plastic foam is **flammable,** it should not be left exposed or covered with paneling, but should be covered with fire-resistant gypsum wallboard.

Reflective insulation This has become quite popular, and it is made from reflective foils such as aluminum or polished metallic flakes that adhere to reinforced paper. It provides a shiny, mirror-like surface that bounces radiant heat back from its source. It is available with or without an attached insulation batt, but without the batt it must be placed so there is dead air space on both sides of the reflecting surface. If it touches surrounding materials, reflective insulation will be ineffective as an insulator because it doesn't retard conductive or convected heat.

Cellulose insulation Cellulose insulation is made of defibered newsprint and other materials which have been treated with borax or other chemicals to retard flame spread and rotting. This insulation can be blown into side walls and attic floors, a job which should be done by a professional installer. The do-it-yourself carpenter can hand pour and level cellulose insulation between the joists.

Cellulose fiber is a good choice for insulation if space is tight because it has a somewhat higher resistance to heat-loss and conduction than other kinds of insulation. This means that a thinner layer of cellulose fiber may have the same R-value as a thicker mineral wool blanket or batt. Cellulose is also less costly than mineral wool insulation. Make sure the bag label states that the product meets Federal specifications, has the UL-approved label, or that the manufacturer is a member of the

N.C.I.M.A. (National Cellulose Insulation Manufacturers Association).

Foamed-in-place insulation   Here, urea formaldehyde foam is injected into outside walls through small drilled holes (these holes are later plugged and the siding reapplied.) The foam flows into all the cavities of the wall as well as around pipes, electrical connections, etc. This means that all cracks and crevices will be filled and all air leaks closed. Urea formaldehyde cures into a honeycomb-like material that will not settle or shift. It is fire- and moisture-proof and is an acoustical insulator. This material is more expensive than other forms of insulation, but is highly efficient and has more insulating value than blown-in materials. After the foam is installed, you may notice an odor, but this should disappear in a day or two. Special skills and equipment are necessary for proper installation, so it should only be done by an installer who is certified or franchised by a reputable company dealing with foamed-in-place insulation.

The National Bureau of Standards, however, **warns** against using foam insulation with urea formaldehyde base in attics and ceilings. These foams tend to shrink and disintegrate at attic temperatures. You should be warned, however, that the Massachusetts Department of Health has banned the use of urea formaldehyde insulation in that state because of a "significant correlation between the UFF (urea foam) insulation and certain formaldehyde-linked illnesses, such as respiratory difficulties, skin and eye irritations, headaches and vomiting."

Attic Insulation

Up in the attic, check for inadequate insulation between joists; voids between batts or blankets; improper fill, with gaps around bracing, wiring, or other obstructions; too-thin blanket insulation; or no insulation at all.

If additional insulation is needed, it can be a do-it-yourself project, or it can be turned over to a reputable insulating or building contractor. Expenses pay for themselves in a few years through reduced heating and cooling costs.

Instant batts   A quick and inexpensive way to get instant savings on heating and air-conditioning bills is with **instant attic insulation.** It's a four-inch batt-type insulation made **without a vapor barrier** for use directly over inadequate amounts of ceiling insulation.

Not much can be done—really economically—to add insulation to your walls, but with the "instant" batts, it is easy to increase your ceiling insulation—the most important area, in terms of insulation, in your house. Many houses have just a few inches of poured insulation, or batts only two or three inches thick, in the ceiling. Adding additional batts sized to fit between the joists in an unfloored attic is the simplest way to bring present insulation up to standard.

Why are "instant" batts without a vapor barrier recommended for this job? Placing an additional vapor barrier over an existing layer of insulation would trap moisture, soak the insulation below the barrier, and sharply reduce its insulating value. Hence, the new batts have no vapor barrier and are ready to apply.

The four-inch batts, four feet long, in packages containing ten batts, are available from most lumber dealers. One package of batts for application between joists on sixteen-inch centers will cover fifty square feet of floor space. If the joists are twenty-four inches apart, sinply cut the insulation batt into two equal twenty-four-inch lengths and lay between the joists.

Installing instant insulation   Here's how to install the batts:

• First, using a ruler, **measure** the existing thickness of attic insulation by slipping the ruler between the joist and insulation until it is in contact with the ceiling. If you are measuring blown-in insulation, make sure the insulation is distributed evenly between each pair of joists.

• Next, **place the batt** between the joists directly above the existing insulation. Occasionally, spot-check the total insulation thickness to make sure it meets the six-inch standard.

Since the ceiling area ranks as the greatest single heat-loss area in the house, insulation manufacturers content the new batts are to act as a barrier against the transfer of heat by reducing

the work load required of heating and cooling units, while helping to maintain a constant comfort level.

Insulating an uninsulated attic If you want to add insulation to an unfloored attic, that has none, the job is relatively
easy. Simply lay out rolls of insulation over the joists. If the
attic floor is finished, raise a few floorboards here and there and
rake in loose insulation.

However, do not let the insulation block attic **vents.** Be
sure to keep attic insulation three inches away from **heat-producing equipment** such as recessed light fixtures that protrude
into the attic floor areas from the story below. If you are using
loose-fill insulation, install a baffle to separate the insulation
and the vents along the eaves; also construct a framework around
a light fixture to keep the insulation at a safe distance from the
heat source. If you are installing batts or blankets of insulation,
cut them to fit around these problem areas in the attic.

Always provide at least **two vent openings** placed so that
the air can flow in one end of the attic and out the other.

Treat all **electrical wiring** with care as you insulate. Don't
try to pull it or bend it out of the way.

Tools you will need to do the job correctly

- A sharp, serrated-edge **kitchen knife** to cut blankets
  and batts.
- A metal **ruler,** a short length of board or other straight
  edge to cut along.
- A **rake,** or similar tool, to push or pull blankets or
  batts to the eaves' edge if there isn't much head room.
- **Measuring tape** if you don't use a metal ruler as a
  straight edge.
- **Walk boards.** Take several boards, ten or twelve inches
  wide and about six feet long. (Many a do-it-yourselfer
  has stepped on the top ceiling surface and plunged
  right through, because he didn't use walk boards.)
- **Staple gun** for applying wall insulation. This can be
  rented.
- **Portable light** with an extension cord.

An average-size attic can be fully insulated for about $200.

Wall Insulation

How do you know how much wall insulation you have? To check the amount of wall insulation you will need a ruler, a screwdriver and a flashlight. Turn off the electricity, then remove the cover from a light switch on an outside wall. Shine the flashlight into the space between the switch box and the wall material. If there is insulation there already, it is probably sufficient. However, if you find no insulation there, you might consider installing some.

In cold weather place your hand at several spots on an interior wall and then compare its temperature with that of an exterior wall. If the home is adequately insulated, the exterior wall will feel only slightly cooler.

Figuring your needs for wall insulation   Insulation is available from building supply dealers and lumber yards. To figure out how much you need for any ceiling, wall or floor area, multiply the gross area (not deducting for joints or studs) by 0.9 if framing members are 16 inches apart, or by 0.94 if framing members are 24 inches apart. Example: 1,000 square feet of ceiling with joists spaced at 16 inches would require 900 square feet of blankets or batts.

Putting up wall insulation   **Staple** blankets to the sides of wall studs, with staples about eight inches apart. The vapor barrier must face the room side of the wall. Make sure there are no gaps. Blankets should touch the wood framing at the top and bottom of each stud space. If you use more than one piece of blanket for a stud space, butt the pieces together snugly.

Push insulation behind electrical receptacles, switches, piping and ducts. **Stuff cracks** around windows and doors with small pieces of insulation fiber.

Finished walls are, of course, the most expensive part of your house to insulate because they are not easily accessible. However, if you take out a wall or add a room, always take advantage of the situation and insulate any exposed exterior walls.

Basement Wall Insulation

Insulate unfinished basement walls. Nail wood furring strips onto concrete walls and fill the established stud spaces with blanket or batt insulation. (The vapor barrier should face the interior basement area.) Install your wall-finishing material over the insulation and furring.

Another way to insulate basement walls is to attach **rigid insulation boards.** No vapor barrier is needed with this type of installation, since insulation board is impervious to moisture. However, for fire safety you must cover the insulation board with at least one-half-inch of drywall (gypsum wallboard).

Foundation Insulation

Studies show foundation walls can account for 20 percent of the total heat-loss in an otherwise well-insulated building. If you are having a new home built, there is a new product on the market today made of **styrofoam.** This foundation insulation is fastened to the foundation walls before the backfilling, and can cut your winter heating bills by as much as 24 percent. It has a rugged board structure that prevents sagging or settling in the soil, providing resistance to both water and water vapor. This below-ground insulation takes your home another step closer to maximum energy efficiency. The only restriction pertaining to this product is that it is combustible, so it must be properly installed according to instructions.

Floor Insulation

Don't overlook those floors, especially over crawl-spaces, cold basements, and garages. A layer of insulation here will keep heat from escaping. When insulating the underside of floors above unheated areas, use four-inch rolls of blanket insulation with a vapor barrier on one side. This side should be face up against the floor.

Push the insulation blankets up between the floor joists from below. Hold the blankets in place by stapling **chicken wire** across nails driven in the bottoms of the joists.

Crawl-Space Insulation

You can also control expensive heat-loss through crawl-space areas by installing blanket or batt insulation on crawl-space walls. Fit the batts snugly against each other, letting then extend two feet into the floor area. Then install a **polyethylene vapor barrier** over the entire dirt floor of the crawl-space, tucking the barrier under the insulation batts on the walls. Be sure to overlap the joints of the vapor barrier by at least six inches, or tape the joints. Where the wall meets the floor, the insulation and vapor barrier can be held in place by weighting it down with two-by-fours.

One word of caution. Insulation experts warn against insulating the crawl-space in this manner if you live in a **very cold region** such as Alaska, Maine, or Minnesota. The extreme frost penetration during one of these cold winters can make the foundation heave if the crawl-space walls have been insulated. In this case, it would be wise to insulate the floor area over the crawl-space.

If the base of your house is exposed, as in the case of a mobile home, build a "skirt" around it. Materials successfully used for this skirt are: plywood, fencing, corrugated fiberglass and boards. If you contact your mobile home dealer you will find a ready-made skirt available also.

Duct-Pipe Insulation   has a savings per year of $20 to $160. Why? Because sheet metal heat ducts and copper hot-water supply pipes act as heat exchangers. For example, if a pipe runs through an unheated area like a crawl-space or an unused basement, a substantial amount of heat will be lost through the walls of the duct before it reaches the register.

The solution is duct insulation, available in one- and two-inch thick **fiberglass blankets** which can be cut to fit snugly and secured with duct tape. (Before insulating, be sure to tape all loose joints and split seams in the ducting.)

Asbestos-honeycomb insulation sleeves for a variety of steam and hot-water pipes are also available. Metal band clamps secure them in place.

A roll of duct insulation two inches thick, forty-eight inches wide and seventy-five feet long costs about $70.

Miscellaneous Areas Needing Insulation

Check your **water heater.** You don't want to heat the utility room with your water heater, but you may be doing just that. See if your water heater has extra insulation on the top, bottom and sides. You can also contain the heat that radiates from the unit by wrapping the water heater with an insulating blanket. Easy-to-install kits are widely available.

Don't forget to close and tightly seal spaces where air can leak **between living areas and the attic.** These openings are around loosely fitting attic stairway doors, pulldown stairways, and ceiling fixtures such as lights and fans.

Consider installing an **attic fan** to supplement insulation in summer. The sun can raise attic temperatures as much as forty degrees above outside temperatures. An exhaust fan to expel the hot air helps to cool off your house.

How to Find a Contractor

When a professional insulation contractor is needed, you can find one by asking your local utility company for suggestions, consulting friends and neighbors, or looking in the Yellow Pages of your telephone directory under Insulation Contractors—Cold & Heat. Check his reputation and his work thoroughly before signing a contract.

The contract should contain specifications and cost of the job. Let's say you have decided a contractor should insulate your attic floor by using blowing wool. How can you tell if you are getting R-38 performance, or whatever thermal resistance rating you have decided you want? First of all, be sure you get a **written contract** stating the number of bags of insulation to be used and the R-value to be achieved. Also, check the bag label yourself. A federal government specification requires that each bag of loose-fill mineral wool insulation be labeled with technical information, including the maximum net coverage per bag of that particular insulation for all commonly specified R-values. This coverage figure gives you a means of knowing the minimum number of bags of insulation the contractor should blow into your attic floor to achieve a particular R-value. Multiply the overall square-foot area of your attic floor by .90 or .94, then divide that number by the "maximum net coverage" listed on the label for the R-number you want.

When you talk to a contractor's salesman, ask him to show you the **bag label** for his brand of insulation, and if you don't fully comprehend the label, ask him to explain it to you. Then, when the job is being done, stay home and count the number of bags actually used.

If a contractor uses insulation packed in bags that are not labeled, beware! The quality of his material will be unknown, and there is no way you can determine the effectiveness of the job.

Also, ask the contractor about the **insurance** he carries. Does he have insurance to protect his own men if they are injured on your property? And, will you be covered if one of his men damages your house?

Insulation Tax Credit

Did you know that Congress may help you with the money you spend for insulation, storm windows, etc., by making such investments, up to $2,000, tax-deductible? This is how important the energy-conserving program is to our nation, as a whole. If you decide to insulate your home, before you go ahead, check with a local representative of the Internal Revenue Service to see whether what you plan to do qualifies for this deduction.

As the law now stands, 15 percent of the cost of insulating a home, up to a total cost of $2,000 can be deducted from the federal income tax that would otherwise be due. The maximum credit is $300.

And, this credit may be taken by owners or renters. Construction of the home to which insulation is added must have been "substantially completed" before April 20, 1977. Storm windows and doors, weatherstripping, caulking, energy-saving replacement furnace burners, and other energy-conserving improvements qualify for the tax credit, too.

CHAPTER 11

Control Heating and Cooling via Your Windows and Doors

CAULKING AND WEATHERSTRIPPING around windows and doors can provide an estimated $30 to $70 per year savings on utility bills. In many homes, up to 70 percent of heat-loss is due to outside air coming in through window and door casings and building sills.

Test your windows and doors for air-tightness. Move a lighted candle around the frames and sashes of your windows. If the flame dances around, you need caulking and/or weatherstripping. (If every gas-heated home were properly caulked and weatherstripped, we would save enough natural gas each year to heat about 4 million homes, say the experts.)

Other places to check: all joints between window and door frames and siding; along the bottom edge of siding where it laps the foundation wall; inside the basement where the sill rests on the foundation; outside water-faucet plates and other penetrations of the outside walls; joints between wing extensions and joints between porches and main body of the house; where outside chimney and other masonry joins the house wall.

Caulking

Caulking compound is available in bulk form in a can or in cartridges. Though the bulk is a little cheaper, the cartridge is more convenient to use with a caulking gun. Caulking cord is a soft, putty-like material in cord form that is pressed in place to seal wood or metal windows. It is inexpensive and easily installed, but only temporary. Once it is on, the windows may not be opened.

Caulk cracks around windows and doors and at the **joints** wherever two materials meet. Caulking is very easy to do, and will help keep unwanted outdoor air from entering your home, while preventing the escape of heated or air-conditioned inside air. The caulking should also be checked once a year. Replace any loose, broken or brittle caulk with new material. (For more on caulking, see Chapter 5.)

One word of caution. Do not caulk the **drain holes** at the window sill or in a combination storm window. Sealing the drain holes could cause rain water to build up on the window sill and seep into the house.

Windows

A single pane of glass does not offer much resistance to the flow of heat, and uncomfortable convection currents can get started along an inadequately glazed window. That's why, even if you have been careful to caulk and weatherstrip that window, you may still feel a cold draft near a single-glazed window.

Use **double-glazed or triple-glazed windows,** or install storm windows. In a double-glazed window the space between the two panes of glass slows the transfer of heat through the window areas. In especially cold regions, triple glazing is becoming a more popular way to cut window heat-loss/gain even more. If you are planning some remodeling, be sure to specify double- or triple-glazed windows as part of the job.

Windows with **wood frames** or wood covered with vinyl or aluminum (which makes for lower maintenance) are good choices for cold climates, because wood itself is a good insulator. There are also fewer problems with moisture condensation with wood than with metal windows, unless these have a thermal barrier.

Storm Windows

Commercially-manufactured storm window units—permanent or removable—represent a big investment that really pays off in lower utility bills. Storm windows can save energy in summer as well as in winter. Leave the storm windows on **all year around** to keep warm summer air out and to help save on air-conditioning bills.

Single-pane storms are relatively inexpensive. They clip over the regular window, but once they are in place, the storms are difficult to open from the inside. The most expensive—but also the most trouble-free—storm-windows are **combination storm/screen units.** Both the storm sash and the screen slide in the tracks of a combination unit. And, storm/screens are easy to maintain because they stay in place on your windows all year long.

Buying storm and screen windows   Combination storm and screen windows that are permanently installed come in **double-track or triple-track** models. With both types there is an upper and lower glass panel and screen, which can be moved to any position for ventilation. The glass panels either come out or are hinged for outdoor cleaning, and the screen is self-storing in the frame. Triple-track windows are more expensive than single-pane or double-track, but are easier to manipulate.

Most storm windows today are **aluminum,** because it is strong and almost maintenance-free. They come in plain, anodized, and baked-enamel finishes. The anodized finish has the aluminum look, but it has been treated against rust and corrosion. The baked enamel is available in white and several other colors. Both types are more durable than the plain finish, which has a tendency to pit and corrode, especially near seashore areas. Steel storm windows are also available, but these have to be protected against rust.

To a large extent, **weight** is an indicator of quality in storm windows, because it means heavy gauge aluminum has been used, and, thus, that the frames will be strong and durable. When shopping for aluminum storm windows, look for a certification seal from AAMA (Architectural Aluminum Manufacturers Association) which insures that they have met minimum performance standards.

When buying combination storm windows, examine the **corner joints,** which should be strong and tightly fitted so air and rain cannot enter. Overlapped corner joints are usually better than mitered ones. A properly designed storm window will also have two or three small holes or vents drilled through the frame where it meets the window sill. These **"weep holes"** keep water condensation from collecting on the inside of the windows.

Check the **thickness** of the frame, especially the depth of the tracks at the sides of the window. The deeper they are, the better the quality. Better storm windows will have heavier and thicker weatherstripping, which will help reduce air leakage around the windows. They will also have glass that weighs at least eighteen ounces to the square foot, as compared to fifteen ounces in lower-priced windows. The chance of breakage is less with the heavier glass, and there is less optical distortion.

Look at the quality of the **hardware** on the storm windows. All locks and hinges should move freely and be easy to handle. Metal catches that hold the window or screen in place are generally better than plastic catches. And, the more notches for holding the window and screen in various positions, the better the window.

After the windows have been installed, see that both windows and screens move smoothly in the sash and close tightly. Also, be sure there is a tightly caulked **seal** around the edge of the storm window where it meets the house, as any leaks there will cut down on efficiency.

Make your own storm windows   If you don't mind obscuring the view, you can make your own storm windows from six mm. thick **polyethylene.** These sheets of clear plastic attached to windows will also convert windswept rooms into draft-free ones. This material, cut to size with shears, can be easily and quickly tacked or stapled to an average-size window for about 60 cents per square foot. People who rent may prefer this inexpensive yet effective and durable means of insulating which can bring about a 30-percent saving on fuel. Although these temporary storm windows are not costly, they must be reinstalled every year or so because they are not very durable.

Replacing Broken Window Glass

All broken windows should be repaired before winter descends. Don a pair of gloves and use pliers to remove the broken glass. Also wear some kind of eye protection. Use a stiff putty knife to remove all of the putty from around the pane. Under it you will find the glazier's points—tiny metal points that hold the glass against the sash. Remove these with a screwdriver or pliers and thoroughly clean the rabbet—the wood recess in which the pane sits. Use a small chisel to cut out any putty that does not come out easily. However, take care not to gouge the wood of the sash while doing this.

Paint the rabbet with an oil-base primer to seal the sash and to keep it from absorbing oil or moisture from the putty or glazing compound.

Before setting the new glass in place, many pros first run a thin layer of compound around the rabbet to cushion the glass and to prevent interior condensation from penetrating between the glass and the sash. This **back-puttying** step is the "right" way, but it makes it quite difficult for beginners to fit a large pane of glass tightly against the rabbet. You can actually skip this step and still do a satisfactory job.

Measure the glass opening exactly and deduct $\frac{1}{16}$ inch from the dimensions for a small pane, $\frac{1}{8}$ inch from a large one (to allow for the pane's sitting down in the rabbet). Cut your own glass or get it done at a hardware store or lumber yard.

Lay the glass in the opening, then set the glazier's points (if you are working with steel sash, you will need steel sash "clips" instead of glazier's points). With a screwdriver, push glazier's points, available at the hardware store or lumber yard where you purchased the glass, into the sash to the left and the right of each corner, and every other four inches along the glass. Now you are ready to fill in the grooves or corner between the glass and the front edges of the frame with putty or **glazing compound.**

Apply glazing compound evenly around the glass, holding your putty knife at a forty-five-degree angle to form a smooth wedge of compound joining the glass and sash. Try to remove excess material that oozes onto the sash and glass without disturbing the smooth compound. Give the compound several days to harden and dry before you paint it. Be sure to join the glass and the compound by running paint about $\frac{1}{16}$-inch onto the glass.

Awnings and Sun Screens

If you are installing awnings, try to place them at heights that will not block the sun's rays in winter, but which will also give you protection from high sun in summer. Or, instead of awnings, consider attaching **sun screens** to the outside of your windows. The cost is nominal and the screens are effective in winter and summer.

Window Screens

Window screens are both simple and inexpensive to repair. You may patch a screen with a scrap of matching wire mesh if the scrap is in good condition. Or, screening fabric is available in rolls or in patches for both major and minor repairs.

How to replace damaged screens   If the screen is badly corroded or torn in several places, it will have to be replaced. There are several different kinds of metal and plastic screening material on the market today, but one of the simplest to install is fiberglass. It won't dent or crease, is easy to stretch tightly by hand, and cuts easily with household scissors. Once it is installed, it won't shrink, stretch or corrode.

To cover a wooden-framed screen, first pry off the moldings around the screening, being very careful not to damage them, as they will have to be put back on. Remove old tacks, staples and screen wire. Cut new screen mesh to size, allowing about an extra half-inch of material on all sides beyond the outside of the molding. This makes a hem that will be folded under the molding to insure a better grip with nailing or stapling.

A **staple gun** is the speediest tool for tacking down the new screen, as it requires only one hand for stapling while the other hand stretches the screen. Start on the long side, tacking or stapling the mesh into each corner, and allow the excess half-inch to extend out beyond the molding. Fold over the excess material and staple or tack through the double thickness at one-inch intervals, beginning in the center. Keep the material stretched tightly so it doesn't bunch. Repeat on the other long side of the screen. When both sides are secure, fasten screening in the same manner to both short sides of the screen, making sure the mesh is pulled tightly in all directions. Replace molding over edges of screening.

With **metal frames,** the wire mesh is held in place with a plastic bedding strip that fits into the groove around the edge of the screen frame and is pressed down on top of the screening so that the mesh is clamped tightly into the groove. No staples or tacks are needed. Begin the job by laying the frame flat on a table. Then pry out the existing plastic bedding strips, and pull out the old screen wire. Cut new screening so that it is as large as the outer side of the screen frame. Lay screening over the top of the frame. Trim off corners at a forty-five-degree angle, making sure that the line of cut leaves enough material in each corner to reach into the bedding strip groove. Lay the bedding strip for one of the long sides on top of the screening, and use a thick-bladed screwdriver to tap the bedding into its groove. Repeat on the opposite long side, pulling the mesh tightly as the bedding strip is forced into its groove. Tap the strips into place in the same manner on the short sides. Trim off the excess screening material with a sharp knife or razor blade on the outside of the groove.

Patching a screen   Cut a rectangular opening in the damaged screen just slightly larger than the puncture or tear. Cut a new rectangular screening patch two inches larger than the trimmed hole. Remove the three outside wires on all four sides of the patch to make raveled edges of about one-half inch. Bend the ends of these wires over a block or edge of a ruler to form prongs that will secure the patch to the screen. Place the patch over the hole from the outside, and push the prongs through the screen. On the inside, bend down the ends of the wires toward the center of the hole. It helps to have someone on the outside to press against the patch while you are working from the inside.

If you have very small holes in your screening, they can be mended by stitching back and forth with a fine wire and a large darning needle.

Insulated Draperies

Most insulated draperies have a layer of white acrylic foam bonded to the fabric back. Some also have an additional layer of black acrylic foam to block out the light. These draperies help keep out winter drafts and summer sunlight better than regular lined or unlined draperies. And they have a significant effect on the temperature of the room if the windows (or sliding glass doors) are a major part of an exterior wall.

Don't forget to **close those draperies** against the cold. Open them only when the sun's warm on that window. Recent tests also show that an ordinary window shade cuts 24 to 31 percent of the heat-loss through window glass.

There is also a new **window screen** made of vinyl-coated fiberglass. The screen stops some 75 percent of the sun's heat in summer, and helps repel chilling winds in winter.

There is also a metalized transparent **plastic film** that can be applied to window glass. The purpose of this material is to reduce the sun's heat and eliminate glare without blotting out light. Besides reducing cooling costs in summer, it works in the opposite fashion in the winter, reflecting heat back into the room. This reflective film is more-or-less permanent, although it can be peeled off if desired. The Yellow Pages of your telephone directory lists dealer-installers under Glass Coating and Tinting. There are new do-it-yourself versions of this film— thinner, easier-to-handle, adhesive-backed sheets that only need to be moistened with water to stick to windows. The material is available in some department stores and home centers in rolls of various widths, lengths, and colors.

Weatherstripping Doors

Seal the **thresholds** to control air infiltration. Don't let cool breezes sneak in under your doors. When it is time to replace a worn threshold, surround the new threshold in caulking to seal out unwanted outdoor air.

If cold air pours under your door, you may need a new threshold. There is an aluminum one with vinyl flaps that makes a tight seal against the door. Such thresholds come pre-packaged with screws and instructions. However, you will need a hacksaw to cut the strip to size.

There is still another way to stop under-door drafts. Take the door down and screw a special **aluminum/vinyl channel** to its bottom. Or, door-bottom drafts can also be stopped by installing "vinyl bulb" thresholds. Both channel and bulb weatherstripping require door removal and fitting for installation. Drafty thresholds can also be plugged easily and inexpensively with simple **"sweeps"** that fasten to the outside of the bottom of the door.

Thin **spring metal** makes a highly durable, and invisible, seal for doors and most windows. However, installation is difficult and somewhat exacting. Metal J-strips for doors and casement windows call for precise alignment, but provide durability and an excellent seal.

Metal plates with a rubber or vinyl insert, screwed into the door's threshold, or a rubber or felt strip attached near the bottom of the door, will also seal out drafts and keep in the heat.

Storm Doors

A storm door cuts air infiltration by establishing an insulating air space between the outdoors and the entry door. A storm door not only keeps out winter winds; it is a valuable insulator all year around. If you live in a region where air-conditioning is a must, you will use your storm door all summer, too, to protect your air-conditioned house from blasts of hot, humid air.

Security and appearance are often the main criteria when selecting a prime entrance storm door, but you can also get a good-looking door with energy-saving features. Several companies make handsome steel doors with a core of **polyurethane foam** that insulates against cold or heat. In addition, the doors have a thermal barrier of wood or vinyl, and magnetic weatherstripping that seals the door tightly. Another feature is an adjustable sill to maintain a tight seal against the weatherstripping on the bottom of the door. Doors with windows come with tempered, shatter-proof insulating glass.

Storm doors are also available with interchangeable screen and glass inserts or with self-storing **screens,** so they can be . used for ventilation during the summer.

Many of the same **quality-construction guidelines** for storm windows also apply to storm doors. Look for structural strength, tight joints, good weatherstripping. Look for the certified seal from the AAMA on aluminum storm doors. For safety's sake, the glass should always be tempered against shattering (this is required by law in some states). Because the door closures and hinges are subject to stress, they should be strong and substantial. A cylindrical air-pressure door closure is desirable.

After the door is installed, check that it operates smoothly and closes tightly. Also, remove and replace the exchangeable panels so you know they fit properly.

Installing Weatherstripping

The Edison Electric Institute says, "Weatherstripping and caulking around windows is a must. If they are leaking air, you may be increasing your heating bills by 15 to 30 percent."

A front door that is not weatherstripped is like a six-inch hole in the exterior wall of your house! And, insulation won't stop the loss of heat through cracks around your doors and windows. You must plug those leaks.

Weatherstripping is an inexpensive but very effective way to stop air leaks as is caulking, and any do-it-yourselfer can easily install it. **Exterior windows and doors** should be weatherstripped, as should doors between heated and unheated spaces in your home. (An example of this would be the kitchen door that opens into an unheated garage, or hallway door that leads to the basement stairwell.) Remember to go over all weatherstripping and check its condition before each season, so you can replace any that is not in A-1 shape.

To install weatherstripping effectively, you must first **find the air leaks.** Use a hand-held hair dryer and have someone inside tell you where the gusts come through when you run the dryer around the door frame. Caulk those spots.

Kinds of weatherstripping   For weatherstripping you can use **foam-backed or vinyl-backed aluminum strips,** or easy-to-apply **foam rubber strips** with an adhesive backing. The first two nail in place; to install foam strips, open the door and clean the jamb and the leak points. When it is dry, peel off the foam's backing and press the strips on. Use your hair dryer for a few minutes to be sure the drying is complete.

Many types of weatherstripping are available today, making the choice of materials a complex decision. Some are rigid. Some are flexible. All will do a good job of sealing if package instructions are followed and they are put on properly. Here is a run-down on what is available—with pros and cons:

Kinds of Weatherstripping

Description	Pros	Cons
Soft, puttylike material in cord form that you press in place to seal either wood or metal windows.	Inexpensive and easily installed.	Temporary — the windows cannot be opened.

Description	Pros	Cons
Felt encased in metal strips around door or window framework to seal crevices.	Durable, works well on doors and windows that are opened often.	Not very attractive. Must be fastened with many small nails.
Strips of U-shaped vinyl that slip over edge of metal casement windows.	Inexpensive, permanent. Easy to install. Window will open and close.	Works only if window fits perfectly, or is not bent or distorted.
Clear, self-adhering tape that seals wood or metal windows.	Inexpensive and easy, just press half the tape on the window and half on the frame.	While it is on, window cannot be opened.
Adhesive-backed foam strips that adhere to framework where the door or window meets it — works equally well on metal or wood.	Easily applied.	Not durable, as it wears where parts slide against it.
Strips of felt that are tacked or stapled on to plug the gap where the door or window meets the framework.	Inexpensive and easy to install if stapled.	Temporary, will deteriorate in time. May not be successfully used where parts slide against it.
Thick strips of foam to stuff into gaps between window sash and frame.	Easily installed, good for sealing extra-large crevices, such as where the sash won't close completely.	Temporary.

Description	Pros	Cons
Spring-activated strips of bronze that are nailed around wood door frame, creating a tight seal when door is closed.	Permanent weatherstripping.	Difficult to install. Door must be planed before installation.
Vinyl stripping with lip that nails to door or window frame to seal, or replaces weatherstrip in factory-weatherstripped windows.	Durable and inexpensive. Slips easily into place on factory windows.	Many nails are needed to install on standard windows.
Cut-to-size to form a standard door stop — vinyl on edge seals crevice between door and frame.	Permanent, easily installed, attractive, can be painted. Door-closing noise is deadened.	None.
Vinyl gasket to seal space between floor and closed door.	Effective if properly installed.	Door bottom must be planed precisely.
Molded double-lip strip nailed to bottom of garage door to seal door/floor gap.	Permanent weatherstripping — cuts out drafts inside garage and cushions shock of door closing.	None.
Aluminum strips with front edge of vinyl, installed directly to face of standard door stop.	Extremely durable, easily installed, attractive, can be painted. Door-closing noise is deadened.	None.

Description	Pros	Cons
Aluminum base screws to bottom of wood door — vinyl upper part seals the door/floor crevice.	Extremely durable.	None.
Interlock installed on door bottom to seal when door is closed.	Easier to install than some — available in various sizes to fit under any door.	Interlock parts can be troublesome, catching on rugs and carpeting as door moves.

How Much Can You Afford to Spend on Weatherstripping?

Only you can answer this question, but weatherstripping will probably cost about $4 per window and $8 per door. Caulking will cost approximately $3 per window, $4 per door. You can figure on spending about $15 per single pane glass storm window (if you install it yourself) and $40 per combination storm window (installation included.) Check local costs in your area.

You can easily recover the cost of weatherstripping and caulking in one season's savings on heating bills. Although you won't get your money back the first season with other storm window investments, you can look forward to a 15-percent savings on heating bills once the storm windows are in place.

You might think twice before installing glass or combination storm windows if you anticipate moving before the end of their six-year payback period; but keep in mind that energy investments you make now can affect your home's salability. Energy-efficient homes carry bigger price tags than homes with both a history and a future of high fuel bills.

CHAPTER 12

Cold-Weather Check-List for Your Home and Heating Units

Service Your Furnace

Make sure your heating system is working properly, and consider making energy-saving modifications, of which there are many available today. For instance:

For gas heat The best system is a properly-sized furnace with an **electronic ignition.** An **automatic flue damper** should also be considered. Ask your gas utility how your present system measures up, and how much money you might save by retrofitting improvements on your present system or buying a new one.

For oil heat Have the furnace serviced once a year, and have the service person check to see if the **firing rate** is correct. Most oil furnaces are over-fired, as we have seen in Chapter 5.

If you have a forced-air heating system Clean or replace the **filters** at least once a month. Check the **ductwork** for air leaks, especially around joints, and repair them with duct tape and/or caulking. Exposed ductwork should be insulated, as suggested in an earlier chapter.

Service your heating and cooling equipment annually. A once-a-year **cleaning and adjustment** will keep the system performing at peak efficiency, and you will get the most for every fuel dollar you spend. Uncleaned equipment may not admit sufficient air, and will not burn the fuel completely.

About 60 percent of the nation's home-heating systems need cleaning, lubrication or adjustment, builders say.

Throughout the year, help keep your heating/cooling system humming by changing or cleaning **filters** regularly. More sophisticated adjustments should be made by a serviceman.

Thermostat Settings

Most healthy people including infants, can cope well with indoor temperatures in the sixty- to seventy-degree range. Elderly persons should consult their physicians before **lowering the thermostat** drastically, however.

Don't lower the temperature so much that you have to use enormous amounts of energy to reheat the house to a comfortable temperature. If you are going to be away for a weekend, or longer holiday, the lower temperatures are recommended.

Close Off Sources of Heat Loss

Check to be sure your fireplace is not wasting energy If your fireplace uses heated indoor air for combustion, you are wasting energy. **Vent** the fireplace to use outdoor air for combustion so that indoor air you have paid to heat won't go up the chimney.

If you do not already have them, outfit the fireplace with heat-resistant **glass doors.** As noted in Chapter 6, the glass doors will allow the fire's heat to radiate into the room, but will reduce the draw of furnace-heated indoor air for the fire's combustion. When it is time to extinguish the fire at the end of the evening, just leave the fireplace damper open and close the glass doors. With the fireplace thus tightly sealed, the fire can burn itself out without smoking up the house, and your household heat won't be drawn out through the chimney during the night.

Close unheated rooms   If you have a conventional heating system, close off unoccupied rooms and shut their heat vents. But, if you have a heat pump system, leave the vents open.

Clean radiators   If you have radiators, dust or vacuum them frequently. Dirt and grime can impede heat-flow. If radiators need a painting, use flat paint rather than glossy paint.

Remove fans and air-conditioners   Remove those window appliances—fans and air conditioners—that will not be used during the winter. If you do not wish to remove them, cover them on the inside to help keep cold air from seeping into the room.

Check for Trouble Spots

Fix roof leaks   Leaks in roofs are aggravated in winter by ice and snow, especially where roofs abut walls or around chimney flashing. Flashing, the metal strips which extend under the roofing, can be sealed with asphalt roofing cement. If shingles on the roof are damaged, a flat piece of aluminum, copper or galvanized sheeting can be cut for the area affected and driven under the shingle, as a temporary measure.

On **asphalt-composition roofs,** the shingles can be replaced by raising the lower edge of the shingle above the missing or damaged one, inserting a new shingle, and nailing it at the top. If the shingles are not the self-sealing type, put a strip of roofing cement on the under edge of both the new shingle and the one above. If a roof is not insulated, the point of a leak often can be spotted from the attic by noticing a drip or a pinpoint of light. Push a thin wire through the hole from the attic side to help find the leak when you are on the roof.

Clean Gutters   Good drainage in gutters is important to prevent overflow, which can cause icicles to form. Gutters should be cleaned, and the downspouts and leaders checked for obstructions. Remove leaves, twigs or branches from gutters and downspouts. Items like tennis balls can cause complete blockage.

Small **holes** in metal gutters and downspouts can be repaired with weatherproof, self-adhesive aluminum foil tape or asphalt roofing cement. For larger holes or tears, cut a piece of aluminum sheeting adequate to cover the hole and cement it in place with aluminum cement. When it is in place, cement over the edges of the new piece.

Gutters should be pitched to **drain easily,** and the hangers, which secure them to the house, should be checked to see if they are fastened properly. There should be a hanger about every three feet to prevent sag, and if a gutter is sagging, additional hangers can be used to get the proper alignment. Paint **wooden gutters** on the inside to prevent rotting. Cracks or holes can be filled with weatherproof caulking before painting.

Check the house foundation   Masonry foundations are particularly vulnerable to ice damage. Crumbling mortar between brick or stones should be removed and new mortar put into the crack. Cracks in cement foundations should be filled with patching cement to avoid ice damage.

Check the chimney   Check the chimney for cracks and breaks. If there is a protective screen at the top of the chimney, remove anything caught in it that might block smoke.

Check potential sources of damage   Cut off any weak tree branches that, if covered with heavy snow or ice, might snap and damage the roof or break windows.

Prepare for Possible Winter Emergencies

Have an emergency heating source and a supply of fuel. Fireplaces, wood stoves, or coal stoves all can serve in emergencies with proper ventilation. Also have the following **emergency items** on hand:

- Extra food and water supplies;
- Transistor radio with extra batteries;

- Flashlight with extra batteries;
- First-aid supplies;
- Extra blankets or sleeping bags.

On an everyday level, pay attention to saving energy. And, now that all your winterizing has been done for your home and property, sit back and look forward to a warm and cozy wintertime!

CHAPTER 13

Light up Your Life
(Saving with Lights and Lamps)

LIGHTING CONSUMES 11 to 16 percent of all electricity used in the home. And, the experts tell us that conserving lighting energy in the home could slash United States energy consumption about 50 million kilowatt-hours of electricity per day (enough to light about 16 million homes.) There is no need to go on living in the dark ages when a bright new world of lights and lighting fixtures is out there waiting for us, and when lights are as easily accessible as any other home furnishing.

How much lighting should we have? According to lighting experts, we should allow one watt of incandescent light for every square foot of floor space, or up to three watts per square foot if your ceilings are extra high and the walls are dark.

However, instead of having all the lights on in the rooms we are using, most of us try to have only the ones that are necessary. For instance, if two people are reading in the living room, two reading lights are really all the light needed. Ceiling and side lights are just a waste of electrical energy, and we must form the habit of not using them unless necessary. As soon as we "catch" the habit, our electric bills will be less, and we will be cooperating during a continuing national energy crisis. Even the young people in your home will catch onto the idea of saving light once they are reminded to have no lights burning in unoccupied rooms.

The cost of necessary lighting? If you leave on 400 watts for only four hours a day, every day, it will cost you, in a month $8.41 at 7¢ per kilowatt-hour!

132

Fluorescent Lighting

Fluorescent lighting is three times as **energy-efficient** as incandescent lighting. Fluorescent lighting bulbs also last longer than incandescents, and it is now possible to buy them to fit into incandescent light sockets. And, fluorescent lighting offers better light distribution than incandescent, thereby cutting down on glare.

Since a fluorescent tube gives more light and uses less current than an incandescent bulb, it pays to install fluorescent lighting wherever possible. Conserva-A-Lite and Killer Watt are adapters for installing fluorescent tubes in fixtures designed for incandescent bulbs. You can get these adapters at electrical supply houses and many hardware stores.

Fluorescent lights are especially good for rooms where you need strong light, such as in the kitchen, bathroom, sewing and utility rooms. These lights, set under kitchen cabinets or over countertops, are pleasant and energy-efficient.

Most fixtures for fluorescent lights have plastic covers. These may tend to yellow and crack if they are made of polystyrene. The better **acrylic covers** are more durable, and are worth the extra investment, since these fluorescent tubes last fifteen to twenty times longer than incandescent bulbs.

You will find fluorescent lights give off three to four times as much light per watt of electricity used as incandescent lamps do. One 40-watt fluorescent light gives more light than three 60-watt incandescent bulbs (annual savings can be as much as $10 in this one instance.) And, a 40-watt fluorescent lamp gives off approximately 80 lumens per watt, while a 60-watt incandescent light gives off only 14.7 lumens per watt.

Does it pay to **switch lights on and off** for short periods? Switching lights on and off does not use enough electricity to cause concern. But, the life of a fluorescent tube is shortened by about two hours each time it is turned off and restarted (though there is little effect on the longevity of an ordinary bulb). Thus, where your fluorescent lights are not in use, it is better to leave the fixtures on for an hour than to switch them on and off.

Facts about Light Bulbs

Before you buy the most popular brand of light bulb, or the one on sale, take the following tips on getting the most value for your money, and a longer-lasting bulb, too:

• **Read the package** carefully. Federal law requires that bulb life be clearly printed on the package, along with the amount of light the bulb produces.

• Compare the bulb's **life-rating** with other bulbs, since the life of the bulb may be more important than wattage or the amount of energy the bulb consumes, especially when installed in hard-to-get-at places.

• Understand what **long life** really means. "Twice the life" may mean that a bulb that used to last 750 hours now lasts 1,500 hours, but a 2,500-hour life is still better than a 1,500-hour life.

• Check the fine print on the package. For example, if the bulb has a **brass base** instead of an aluminum one, it is less likely to freeze (or become stuck) in its socket. Also check for warranty protection. If a manufacturer's defect shows up, your light-bulb investment should be protected.

• The U.S. Department of Agriculture says that since two different brands of 100-watt light bulbs may give different amounts of light while using the same amount of electricity, examine the **lumen rating** on the package of bulbs before you buy. The higher the lumen rating, the more light you will get for your money.

• In spite of what you hear about buying **"lifetime" light bulbs,** the experts say the best buy for general use is the standard light bulb, which uses less electric power and converts more of it into light. The standard 100-watt frosted bulb, costing about 35 cents, burns for an average of 750 hours—the equivalent of about 31 days around the clock.

• Remember that a single 100-watt bulb sheds half again as much light as four 25-watt bulbs, so buy **higher-wattage bulbs** to economize, without going over the rating of the fixture. (However, remember the rating of the fixture is very important. Four 100-watt bulbs in a ceiling fixture that is supposed to have a maximum of 60-watt bulbs can make your ceiling crackle and scorch.)

• **Reflector bulbs** used over work areas will focus the light downward, so that less energy is wasted.

• Use a **130-volt bulb** in hard-to-reach sockets, such as stairwells. "Edison" bulbs from your power company provide you with 120 volts. Consequently, the 130-volt bulb will never get hot, and its life will be extended considerably. However, a long-lived bulb gives you less light, so don't use it for reading.

Pros and Cons of the Longer-Life Bulbs

Today's longer-life bulbs provide only a little **less light** than standard bulbs and last up to **four times longer.** For instance, GE and Sylvania make Soft White Plus bulbs rated at 1,500 hours. Westinghouse has the Super Bulb, filled with krypton gas, rated at 3,000 hours; Duro-Lite's X2500 is rated at 2,500 hours. Sylvania's SuperSaver and Duro-Lite's Watt-Saver, both filled with krypton, are also rated at 2,500 hours (and in addition, use about 8 percent less electricity than standard bulbs.) And, while these bulbs will last as much as four times longer than conventional bulbs, the price does not increase proportionately.

Buying Lamps and Ceiling Fixtures

Table lamps For the right spread of light, table lamps should measure thirty-eight to forty-two inches from shade bottom to the floor. For good **reading,** use at least 150-watt bulbs, and set the lamp so it is in line with your shoulder when you are seated beside it. Translucent shades give off more light, whereas opaque shades direct the light sharply up and down.

Buy table lamps in scale to the table they will sit upon. As a rule, the lamp shade should be two-thirds as tall as the lamp.

Floor lamps   Floor lamps should measure from forty to forty-nine inches from the base to the shade bottom. The rule of thumb here is that they be tall enough so you should never look down onto the bulbs when you are standing, yet short enough so you never must look up into its glare when you are sitting. For comfortable reading, three 100-watt bulbs are recommended for floor lamps.

Desk study lamps   Studies really call for the proper light. In order to avoid eyestrain, you must have overall illumination in the room. The desk study lamp should measure fifteen inches from the shade bottom to the desk top. A white shade and a frosted bulb up to 250 watts is recommended for this area. Place the desk lamp fifteen inches to the right of your work center, or desk, if you are right-handed, and twelve inches from the desk front. The experts also tell us to avoid high-intensity lamps in this area unless they are supplemented with overall illumination. And, avoid a shiny, glare-prone desk surface.

Night reading bed lamps   If you love reading in bed, the proper reading light is important to you. The shade bottom should be at eye level when you are lying back, and the lamp base should be about twenty-two inches to the right or left of your shoulder. A flexible, three-way bulb (from 50 up to 200 watts) is recommended.

Try 50-watt reflector floodlights in directional lamps such as pole and spot lamps. These flood lights provide about the same amount of light as the standard 100-watt bulbs but at half the wattage.

Chandeliers and hanging lights   Exposed bulbs should be low in wattage to avoid glare and always hung above eye level, between thirty- and thirty-six inches above a dining room table, for instance.

Ceiling fixtures   Today, ceiling fixtures include recessed downlights in the ceiling as well as surface-mounted track lights. The track lights are especially good for focusing light in a specific area—kitchen or work counters, a wall with art works, etc. Several track lighting kits are available to the do-it-yourselfer.

You simply set the track into the ceiling and run the cord down an inconspicuous corner to a plug or outlet.

Dimmer Switches

The dimmer switch installed on incandescent lighting allows you to "tune down" the light level and wattage, saving up to 50 *percent* on electricity. All you need is a screwdriver to install it.

Use three-way bulbs and dimmer switches wherever possible, so that you can vary the intensity of the lighting depending on your needs.

Dimmer switches actually make your incandescent **light bulbs last longer,** too. These switches—which can be purchased for under $15—are installed in existing switch boxes by turning off the power and connecting two terminal wires. When the dimmer cuts voltage by one-half, dimmed bulbs will use only one-third of their normal wattage.

How to install   Go to your fuse box or circuit breaker and shut off the circuit that feeds the switch you will be working on. This is very important, for obvious reasons!

First, remove the screws holding the switch plate, then the inner screws holding the switch to the wall. Pull the switch out and loosen the screws that hold the wires. One set of wires is usually black, and one set white (or red). Remember which went to which screws, and hook the right wires around the screws on the new switch.

Fold the wires back in the box. Screw the switch to the wall, and finally replace the switch plate—and the job is done!

Timers

If you use lighting to indicate that someone is home, in order to discourage burglars, use timers to turn the lights on and off. Don't leave lights burning constantly, which looks phoney to anyone watching your house. Also, control outdoor safety-security lighting with a timer or photocell switch so the lights go off automatically at a pre-set time, or at dawn, rather than burning needlessly into the morning—or even all day, if someone forgets to turn them off.

Outdoor Lighting

Have those decorative outdoor gas lamps turned off unless they are essential to safety, or convert them to electricity. Keeping just six or seven gas lamps burning uses as much natural gas as it takes to heat an average-size home. By turning off one gas lamp, the experts claim we will save about $27 a year in natural gas costs. If you convert one gas lamp to electricity you could save almost $19 per year! Check gas and electric prices in your area to see if these figures pertain to you.

Another replacement for outdoor incandescent lighting would be low-wattage sodium or mercury vapor lamps. They consume less energy for a given light output, and last ten to twenty times longer than incandescent lights.

Miscellaneous Lighting Tips

Replace old, darkened bulbs *before* they burn out. Near the end of their life, light bulbs give out substantially less light. The tungsten filament thins out and vaporized tungsten collects as a black deposit on the inside of the glass, blocking light output. But, the bulb still uses the same number of watts. Since bulbs consume four or more times their initial cost in electrical energy during their lifetime, throwing away a bulb a few hours before it burns out is really no loss.

Always turn **three-way bulbs** down to the lowest lighting level when watching television. You will reduce the glare and use far less energy.

Where fluorescent lamps cannot be used, try the relatively

new watt-saver bulbs. They can save you eight percent of electricity, without loss of light intensity.

You will get brighter light from lower wattage if you keep those light bulbs **dust free** and vacuum the lampshades often. Remember, you can buy 4-watt bulbs to use as night lights, instead of the 7-watt size.

And, why not use 25-watt light bulbs in closets, utility room, basement, attic, garage and other spots where only a minimum amount of light is required?

Another excellent way to stop wasting electricity is to install wall switches with **indicator lights** on circuits where the light might accidentally be left on—in the basement, attic, garage or for outdoor lights.

CHAPTER 14

Saving Energy with Your Auto

AUTOMOBILE ENERGY CONSUMPTION is a large part of our total energy consumption. This means that the potential for saving is proportionally large.

There are more than 100 million registered automobiles in the United States. A car with an average fuel economy of less than 15 miles per gallon travels about 10,000 miles each year and uses well over 650 gallons of gasoline.

Altogether, our private automobiles consume some 70 *billion* gallons of gasoline each year. That's over 6 million barrels a day, or about the amount of petroleum currently being imported into the United States.

Did you know the heat energy of a gallon of gasoline is equivalent to fifteen man-days of labor? One barrel of oil contains heat energy equivalent to the energy of a man at hard labor for two years.

A spokesman for Shell Oil Company says, "If every passenger vehicle in America used only one less gallon of gasoline per week, we would save over 112 million gallons every week. This savings would meet the President's request of U.S. motorists for a 5-percent reduction in gasoline usage. And for many of us, just observing the 55-mile-per-hour speed limit would do it!"

You have to be a very vigilant motorist to keep your car from being a very "fuelish" gulper of today's high-priced gasoline. There are many minor ways your car can waste precious fuel. Some of these losses may sound minimal, but added together they can represent a bundle of dollars spewing from the exhaust pipe of your car.

If your engine needs a **tune-up,** for instance (and it is estimated that about 60 percent of the cars on the road need a tune-up), your fuel costs can be about 10 percent higher than normal, even if your car still seems to be running all right. If the need is acute, and your car is running roughly, it could be burning about 20 percent more fuel than it should. For instance, a sticky carburetor can cut your miles per gallon by about three miles, and a dirty air cleaner, by about one mile.

Proven Ways to Get Better Gasoline Mileage

• The 55-m.p.h. **speed limit** was enacted to help preserve our dwindling petroleum reserves. Obey it and you will do yourself and your country a service. A car traveling at 70 m.p.h. gets only two-thirds the gas economy of a car going 45 m.p.h.

• If you are stuck in a traffic jam, or in one of those endless lines waiting to buy gasoline, or where construction is rampant on the expressway, and don't move from any given spot for more than one minute, the best way to save fuel is to **shut off your engine.** That advice comes from the U. S. Department of Transportation. After 60 seconds in one spot it takes less gasoline to restart the average car than would be wasted if you kept the engine idling while you waited in line.

• When **climbing a hill,** don't floor the accelerator. That wastes gasoline.

• When you are **stopping on a hill,** hold the car in place by using the brake, not the accelerator.

• **Fast getaways** and jumpy starts can burn 50 percent more gasoline than smooth, slow accelerations.

• If you have a **standard shift,** skip a gear whenever conditions permit. A car running in second gear uses about 45 percent more fuel than a car in fourth. The idea is to get into the economical higher gears sooner. Level or downhill starts are good times to skip second and go directly into third. The opportunities for skipping are even greater with four- and five-speed transmissions.

• **In winter,** a car started cold and driven only one mile will lose about 50 percent of the gas mileage it gets when warmed up. But, this *doesn't* mean you should warm your car up in the driveway. (Group trips should be planned so most of your driving is done while the engine is warm.)

• Most urban signal light systems are programmed for an interrupted traffic flow. If you can **gauge your speed** to make all the green lights—not the yellows and reds—miles per gallon will improve.

• Watch for **engine knock.** It can waste fuel and damage your engine. An engine that is badly out of tune can cut mileage by 10 percent or more.

• In **summertime,** parking in shady areas helps prevent the gasoline evaporation that occurs if you park in the hot sun. Your car will stay cooler, too, meaning less gas-eating work for the air-conditioner once you start on your way.

• **Keep track** of your gas mileage. It is a good indicator that your car may need some maintenance work.

Gasoline and Other Fuels

If you want your car to give good performance, don't look for bargains in gas, but stick with the recommendations in your owner's manual. It is a good idea to keep your gas tank full during the winter season, too. A **full tank** lessens the chance of moisture condensation and subsequent freezing of this moisture in the fuel line. In addition, a full tank comes in handy on snow-packed or icy hills, as it provides added weight for better rear-wheel traction.

The experts add, **don't ever carry an extra can of gas** in your trunk. Carrying a five-gallon container of gasoline in your car trunk is the same as transporting 250 pounds of dynamite. Gasoline vapor, which expands to fill the space in which it is contained, can split the seams of an unvented can or plastic container and cause vapor leakage from a vented "safety" can. A lighted match or cigarette, or a collision, is all that is needed to trigger an explosion.

All motorists should also be aware that carrying extra gasoline in your car can affect your car insurance. The Association of Trial Lawyers of America has warned drivers that carrying gasoline cans in their cars could result in their being held liable in a car accident that resulted in fire or an explosion. **Storing a can of gas** in your garage is equally as dangerous. Although most people know that gasoline is highly flammable, many people don't realize the danger of gasoline vapors. If a storage can is left open in a closed garage, a cigarette lit nearby or even a spark from an auto engine can ignite the gasoline vapors and cause an explosion.

Are brands of gasoline really different? The experts say they certainly are. Gasolines are made of over 150 different hydrocarbons. Some **vaporize** at lower temperatures than others. The way they are blended can make a big difference in the "driveability" of a fuel. An unbalanced gasoline blend can cause problems. What kind of problems? "Stall-out when the engine is cold, or hesitation, say the experts." By "hesitation" they mean a condition like the following: You are about to enter an expressway. You step on the accelerator, and there is a hesitation before your engine responds. Before taking this problem to a mechanic, shop around and see if a different brand of gasoline will help.

There are other differences between gasoline brands. The **detergent additive** that keeps deposits from building up inside the carburetor and inlet system of your car's engine is an important difference. These deposits can reduce gasoline mileage considerably. Some gasolines have really effective detergents. Others have little or none.

Did you know that the same brand and grade of gasoline may vary in different areas of the country? This is because gasolines are blended to meet the changing needs of engines in different seasons and at different altitudes.

What kind of gasoline do you buy—**premium or regular?** These terms are not like "Prime" and "Choice" in grading beef. You are not giving your car a great treat by feeding it premium when it does not actually need it. Look in your owner's manual and see what fuel is recommended.

Differences in gasoline brands result primarily from refining processes used by the various oil companies. These differences needn't concern you, because they don't affect engine performance. National brands also differ in the use of additives. Most brands contain additives to reduce accumulations of gummy dirt and carbon in spark plugs, cylinders and pistons; to prevent unwanted chemical reactions between the gasoline and metal; and to keep the fuel lines and carburetors clean. During the winter months, gasoline sold in colder climates also contains anti-icers to keep fuel lines from freezing. If you are interested in learning the specific additives of any national brand, ask the service station manager. You will most likely find several brands that will keep your engine performing satisfactorily.

The next step is to shop for **price**. Discount brands may lack one or more of the additives found in most national brands. Some of this gasoline is refined specifically for the discount chains by one of the major oil companies, and several competing discount stations may sell the same gasoline. Again, ask the station manager to identify his supplier and list the gasoline's additives.

What about those **off-brands** of gasoline? Did you know that most independent gas stations get their fuel as surplus from the same sources as the stations selling advertised brands? They are simply able to sell it for less. But now that the surplus disappears periodically, many independent stations are curtailing their hours or closing down completely. If you can still find "off-brand" gasoline, by all means, buy it.

Unleaded gasoline   Authorities tell us automobiles account for from 60 to 80 percent of all *air pollution*. For every mile the average car travels, it spews out 11 grams of hydrocarbons (mostly unburned fuel), 80 grams of carbon monoxide (a colorless, odorless, poisonous gas), 4 to 6 grams of nitrogen oxide (another poison gas), and .1 to .4 grams of particulates (solid specks of debris so small they can be inhaled). Lead—which is a poison—is found only in those specks, not in the gases. The rest of the specks are rust, soot, engine debris, etc. According to the World Health Organization, and the American Medical Association, lead from auto exhaust does not constitute a danger to people. However, scientists say that lead particles in the atmosphere are not entirely harmless. A grain of lead can become a full-fledged blob of smog when combined with other particles in our atmosphere.

Many steady users of lead free gas have long boasted that their cars hummed along to a ripe old age with remarkably **few engine troubles.** In a car that uses only leaded gas, the spark plugs have to be replaced every twenty thousand miles or so. Using unleaded gas, they will last about thirty-three thousand miles. Mufflers also last longer in the absence of lead. So do the tail pipes. In addition, unleaded gas will allow increased time between oil changes, greater intervals between tune-ups, and it will provide greater rust protection.

The main difference between leaded and unleaded gasolines is that unleaded gasolines take more crude oil to make, and generally **cost more.** And, you can expect the price gap between unleaded and leaded gasolines to widen even further in the future because refiners will be using more exotic additives and refining processes to produce fuels that pollute less.

If your car requires unleaded gasoline, by all means use it. Using leaded gasoline in a car requiring unleaded spoils the catalytic converter, and it is against the law. Strike a balance between price and performance. The octane and driveability needs of your car may make it worthwhile to use a premium grade of gasoline. However, don't pay the difference in price unless your car really needs the added octane. If it runs well on a lower octane, or on a lower-cost grade of gasoline, use it and pocket the savings.

And John Bame, expert in gasoline and gasoline technology with the Shell Corporation, says, "We estimate that about 70 percent of the cars that require unleaded gas will run satisfactorily on regular unleaded. The rest, however, will knock, and some will also run on after the key is turned off. They should try a higher octane unleaded.

"The thing to remember," he adds, "is if premium unleaded helps your car run better, buy it. Otherwise, don't. It costs more to make and uses more energy in the refining process. So stick to a lower octane unleaded if you possibly can."

Octane requirements   If your owner's manual doesn't provide information on the level of octane requirements, the Department of Transportation suggests you have your car tuned. Fill up with the brand of gasoline you usually buy. Drive a few miles and, after the engine is warmed up, come to a complete stop and then accelerate hard. If there is a knock or ping, switch to the next highest octane on your next fill up and repeat the test. When the knock goes away, you have found the correct octane. If there was no knock, switch to a lower octane and continue switching downward until you notice a knock, indicating that the octane of the *previous* fill up was best for your car.

What is an octane number? The octane number is a **measure of a fuel's ability to resist knocking.** And, the average of the Research and Motor Association ratings is required by law to be posted on each gas pump. This average usually falls between 87 and 95, depending on the particular brand and grade of gasoline. (And, your engine's octane requirements will increase with age as deposits build up.) Thus, a new car that runs well on 87 octane today may require a higher-octane gasoline later.

With the emergence of the **catalytic converter,** most late model cars now require unleaded fuel with a minimum octane level specified by the Environmental Protection Agency. However, a problem has loomed. Many of these cars develop octane requirements higher than the EPA's minimum level. So, gasoline refiners are making more of the higher-octane unleaded gasolines.

The importance of engine knock   That pinging or knocking sound from your engine when you accelerate usually means the fuel-air mixture in the cylinders after ignition by the spark plug isn't burning snoothly all the way across the chamber. With engine knock, after a portion of the fuel burns smoothly, the last part explodes violently. A light, occasional knock won't hurt your car's engine. But heavy, sustained knocking can lead to severe engine damage.

What causes run-on?   When your engine continues to run in a very jerky and unfamiliar way after you turn off the ignition, the problem is called "run-on." There is no spark in the combustion chamber, but the fuel-air mixture enters the chamber and ignites because of the high temperature. The cause could be a gasoline with too low an octane rating. If the condition persists, have your mechanic check it.

Fuel efficiency in different seasons   The experts tell us, "Summertime is the best economy time. Your car requires less power in warmer weather. The difference between 20 degrees and 70 degrees in fuel economy on a 20-miles-per gallon car is about 1.5 miles to the gallon."

What is gasohol?   One bushel of corn can produce 2.6 gallons of ethanol, which when combined with 23.4 gallons of gasoline yields 26 gallons of gasohol.

Across the Farm Belt, "gasoholics" maintain rural Americans can invest in their own economic well-being by choosing higher-priced gasohol over unleaded gasoline. They say everyone wins—grain producers, drivers who say their cars run more smoothly, and consumers who can cut the nation's oil imports by $4.5 billion a year. Too good to be true? We really don't know. At this writing, the jury is still out.

The cruel fact is the price at the pump for gasoline approaches $1.50 a gallon, draining family budgets. The foreign import of crude from OPEC daily drains U.S. dollars, and almost weekly one of the OPEC nations announces another price increase for crude oil.

If all American cars were run on gasohol, we could decrease our independence on OPEC crude by 10 percent of that oil which is distilled into gasoline. That would be a long step toward negating our dependence of OPEC oil, would partially stem the flow of U. S. dollars to the Middle East, and would aid in the battle against domestic inflation.

There are many claims that gasohol is a superior fuel for gasoline. While they are claims of interested groups, they stand well as there seem to be no serious disclaimers. The claims are:

• That 10 percent of alcohol mixed with gasoline raises octane three or four points. The higher octane reduces or eliminates engine knock and ping. Gasohol's higher octane also cuts engine dieseling and "run back."

• Alcohol in gasohol is a de-icer and prevents freeze-ups

in cold winter driving. It is less expensive than commercial de-icers.

- Tests show that after 15,000 miles of driving on gasohol, spark plugs show little or no wear.
- No modifications or adjustments need be made to run on gasohol.
- Gasohol acts as a solvent—a cleaner—in the car's engine. This will cut engine dirt and deposits.
- Gasohol burns cleaner than gasoline and more thoroughly. It will actually burn off some deposits that may have accumulated in the car engine.

And, gasohol today is being sold in more than 800 retail outlets in 28 states, and seems to be going over big! For instance, about 100 Illinois filling stations carry it, mostly in rural downstate areas, and more and more independent dealers are offering it to meet the growing demand.

"It's going over very well," said Len McEnery, district manager for Gas City, which has twenty-three stations on the southwest side of Chicago.

"We swear by the stuff," said John Sweeney, vice president of Sweeny Oil Company in Palatine, Illinois. "We use it in all of our company trucks and cars, and in our personal cars as well. And, our customers keep coming back for more." Sweeney became one of the first Chicago area dealers to offer gasohol in April, and sold about three thousand gallons in the first nine hours. Sales have leveled off since then, but Sweeney still sells his complete supply of gasohol each month.

One of the newest sellers of gasohol is Standard Oil of Indiana, which began test-marketing gasohol at six of its Amoco filling stations in Decatur, Illinois on August 1. Amoco is the only major oil company to make the product available in Illinois. And preliminary results are good, says Paul Collier, Amoco regional vice president. "Sales are approximately at the level of the product replaced (unleaded premium)," even though gasohol costs three or four cents a gallon more, he said. "The customers don't seem to mind at all."

The high price of gasohol reflects the added cost of adding one part of 200-proof alcohol to nine parts of unleaded gas. To keep prices more in line, the government last fall eliminated the **motor fuel tax** on gasohol.

If gasohol continues to rate high marks and governnent support, the Energy Department predicts, production of alcohol for gasohol could increase from a current level of 60 million gallons to 300 million gallons a year by 1982. "Gasohol use would thus reach 3 billion gallons per year, or 3 percent of present gasoline production by 1982," the report concluded.

Diesel fuel   At this point we really don't know. It could be the real savior for the family-size car, or it could go the way of the Edsel and tailfins.

We see some Volkswagen devotees who are willing to pay several thousands of dollars over list to get their hands on a Rabbit with a diesel engine under the hood. However, the federal government looks under the hood at this writing and says that the diesel engine is not clean enough.

The diesel obtains about 25 percent better **fuel economy** than a comparable gasoline engine. It burns less costly diesel fuel, which runs anywhere from 2 cents a gallon less than regular leaded gasoline to 10 cents or more less than lead-free gas.

There are fewer moving parts in a diesel engine than in a gasoline engine, and about all a tune-up entails is changing the oil and filter.

But, the diesel engines also have some problems. One is the fuel. Diesel fuel and home heating oil are just about the same. Thus, a cold winter means more production of home heating oil, causing **diesel fuel shortages.** And, the cost of diesel fuel has also soared, like that of every other type of fuel.

Perhaps the biggest stumbling block to more extensive diesel engine usage is the **emissions law.** Diesel engines emit only small amounts of carbon monoxide or hydrocarbons, but more oxides of nitrogen, or soot than gasoline. The present federal emission standards allow 2 grams of nitrogen oxide a mile, but in 1981 the standard goes to 0.6 grams and in 1983 to 0.2 grams. The auto industry claims the reduction just can't be done.

Even without the fuel and emissions problems, diesels have

some peculiarities. To be frank about it, they **smell.** To ride inside one or behind one is similar to taking up residence near a gas pump. The engines are also **noisy,** because of the diesel engine's combustion process. The air-fuel mix is compressed and then ignited, with the heat generated by glow plugs, which are similar to the coil element on an electric stove.

The diesel is also **sluggish,** partly because the air-fuel mix is about 22 to 1. The diesel, a heavy engine, also requires a pair of heavy-duty batteries to ensure cold weather starts and lots of weight-adding sound insulation to keep the pinging to a minimum within the passenger compartment. (The combustion in a diesel makes a "ping," whereas there is no such noise when the spark plugs ignite the air-fuel mix in a gasoline engine in good condition and using the proper octane level of gasoline.)

During **cold weather** diesel fuel tends to jell at zero, making starts a problem. You might be able to boast fifty miles per gallon, but you surely can't get it when you are standing still. So, if you are interested in a car with a diesel engine, be sure to look into the latest innovations to see if these problems have been corrected.

Buying and Changing Oil

Brands of oil also differ in their additives. Detergents are oil's most vital additive, since they keep combustion by-products—sludge and carbon—and dirt from increasing the friction and wear on moving parts. Other additives in good oils neutralize acids in the engine, inhibit metal corrosion, reduce foaming, which would cause the oil pump to gulp air bubbles, and make the oil even more slippery.

If you buy oil at a store, you should consider only the familiar **national brands.** Of course, these oils will have higher price tags, but the cheap oils generally wear out quickly or are unsuited for even brief use. Likewise, choose a national-brand oil filter.

Also, when buying oil, remember that a **lighter (lower**

numbered) oil will flow more easily and lubricate better when cold than will a heavier, more viscous oil. It also takes less engine power to pump the lighter oil. Thus, a 10W oil allows better gas mileage than a 40W oil, especially in winter. If you feel more comfortable with a heavier oil, on the principle that "heavy oil protects better," then a multigrade oil is the answer. A 10W-40 will be thin when cold, then thicken as the engine warms up and demands extra protection.

Synthetic Oils   What do the experts say about the new synthetic oils? By reducing internal engine friction, they do **promote better mileage.** However, the improvement varies greatly, depending on the condition of your engine, and the cost of the new synthetic oils is high—up to $4 per quart. But, better mileage and prolonged engine life can be worth it.

Further, manufacturers claim their relatively new synthetic oils, made of polyesters or other man-made materials, provide superior lubrication and durability that result in improved gas mileage and easier starts in winter. Check to see if the manufacturer of your car balks at accepting a warranty claim if you use a synthetic oil while the car is under warranty.

Check carefully the pros and cons of using synthetic oils. For instance, some synthetic oils are supposed to last for 15,000 miles and others for 25,000 miles. Currently, prices range about $5 to $6 per quart. Carefully test gas mileage and **calculate your costs** for at least one period of 15,000 or 25,000 miles. Add up the cost of the oil and filter (and labor, if a service station puts the oil in), and compare this to how much the three or more oil changes with regular oil during the period would have cost. You will also have to figure in one or two oil filter changes during the period, as recommended by the synthetic-oil manufacturer, since filters wear out before the synthetic oil. And, when you change the filter you will probably lose about a quart of oil, so add the cost of replacement oil to your tally.

The manufacturers of synthetic oil also claim you will save an average of about 4 percent on gas consumption—so you may have to keep careful records of gas purchases for some time before you can come up with the exact saving.

Tires

Fuel-economy tests show that **radial tires** improve gasoline mileage. A car that gets 15 miles per gallon with conventional tires would get 16 miles per gallon with radials, and would use 166 fewer gallons of gasoline if driven 40,000 miles—the usual guaranteed mileage for radial tires. Radial tires can actually boost your mileage by about 6 percent.

Also, stay away from the popular **wide-track tires** if you want to top mileage. Narrow-tread tires produce less friction and thus less rolling resistance.

And, it is best to buy four radial tires at a time, of the same type (never buy only one pair for the front wheels.) Ask your car or tire dealer if the car can take radials.

Checking your tire pressure Air is one of the few free things left in this world. And, if you run your tires on too little air, they will tend to become hotter, and the tread will wear out prematurely around the edges. Too much air in the tires will cause the tread to wear out sooner than it should in the center of the tire, and you will have less traction. This is important, because when traveling on snowy or wet roads, you need all the traction you can get.

Improper air pressure in the tires also decreases gas mileage and increases wear and the risk of a blowout. Worn-out edges can signal under-inflation, while a worn-out center usually means over-inflation.

First, you must determine what the **recommended pressure** is. You will find it in your owner's manual. Then, the pressure should be checked at least once a month, although you must visually check your tires every time you get into the car to drive. Remember that a radial tire will bulge more than a non-radial.

The recommended tire pressure is for cold tires, when your car hasn't been driven for a few hours. That's why it is best to check the pressure before you start to drive, or before you have driven more than a mile or so. If you check your tires when they are hot, they will be over the recommended pressure if they are correctly inflated), but don't let any of the air out. The increase caused by heat is allowed for in the recommendations.

If you are about to take a long trip, the experts tell us it might be well to add two to four pounds over the recommended pressure. But never exceed the amount shown on the side of the tire.

Also, when checking tire pressure, remember that a ten-degree drop in temperature will reduce tire pressure by one pound. The difference between winter and summer pressures, unless adjusted, can be as much as eight pounds, or could cost you two miles per gallon.

Underinflated tires can cut one or more miles from your car's miles per gallon. And, driving on gravel roads reduces mileage by about 35 percent compared to driving on smoothly-paved roads.

How important is tire rotation? Tires, of course, are very important to the overall energy-saving picture for operating your car. Improper front-end alignment can cut miles per gallon by about one-third mile. Your owner's manual should specify when you should have your tires rotated. However, if you have your wheels switched **every five thousand miles,** you should get equal wear out of your tires. This is not only a good safety measure, but it is good for the pocketbook, as well. If you do a lot of high-speed driving on expressways or tollways, you will find your drive wheels (the rear wheels on most cars) tend to show the most wear. City driving places greater stress on your front wheels.

Wheels that are out of line and tires that are unbalanced not only cut mileage, but all kinds of car problems show their symptoms in uneven tire wear. If your tires seem to shimmy or vibrate at a particular speed, don't just avoid that speed—find out what is wrong. Your tires may be out of balance, or you may be having a suspension-system problem. It will cost you less in the long run if you have the problem taken care of right away, before tire life is shortened. Tire alignment and balance are necessary for smooth riding performance while getting the maximum mileage from your tires.

How can you tell when a tire should be replaced? Many tires today have **built-in indicators.** When the tread depth gets down to one-sixteenth of an inch, a smooth band appears across the tread, and you know it is no longer safe, as traction in inclement weather would be almost nil. If your tires are not wear-banded, you might try this old trick. Stick a **penny** into the tread, and if you can see the top of Abe Lincoln's head in two or more of the grooves, it is time to get new tires.

Winter Car Care

A heated **garage,** of course, is the best winter protection your car can have. But, if yours isn't housed that luxuriously, you can still be sure of quick cold-weather starts with one of the many **heating accessories** on the market today.

However, no heating device is going to help much if your car's engine or electrical system isn't operating properly to begin with. Make sure you have a good battery, that the engine has been tuned (spark plugs, points, and timing should be checked if you have driven more than 15,000 miles since the last checkup). The engine should also be well lubricated with a quality, all-weather winter grade oil.

One of the simplest heating devices is the **freeze plug engine heater.** This is a small metal plug containing an electrical heating element. It is inserted in one of the engine block holes designed to accommodate the core plugs. There it comes in contact with the coolant in the engine block and keeps the coolant heated to about 115 degrees. This heater can be plugged in about an hour before starting your car. In really cold weather, it can be left plugged in overnight.

These freeze plug heaters come in 750-watt size for larger engines and a 400-watt size for smaller engines. They cost from $6 to $10. Inserting the heater can be difficult for most of us, so count on an installation charge of about $10.

Another popular heater is the **oil dipstick heater.** It uses electricity to transfer heat directly to the oil in your crankcase. The heater is a flexible, stainless steel tube with cord attached. This tube is inserted in place of the regular dipstick, then plugged in to be left overnight. Since the heat output is low, best results are obtained when the unit is plugged in while the engine is still warm. These dipstick heaters are reasonably priced under $10.

Batteries can also be kept at full charge in cold weather by warming them with a **battery heater**. Some models fit on the bottom of the battery stand, others clamp to the battery case. To be effective, they must be operated all night. Battery heaters are also priced under $10.

Don't let your **battery run low** in winter, or your chariot will have trouble starting. If you think your battery is low and might need charging, but you aren't quite sure, here is a quick way to find out. If your headlights shine brightly when the engine is running, but appear much dimmer when the car is parked; or if the click of your turn indicator is slower than usual, but picks up when you accelerate, your battery probably needs attention.

The "frosting" that sometimes forms on the top of your car battery is **corrosion** (a mixture of dust, dirt and battery acid). Cleaning it off will prolong the life of your battery. To get rid of this glop, use a stiff-bristled brush and a solution of two tablespoons of baking soda to a cup of warm water. Be careful not to get the solution into the battery or any of the corrosion on yourself. Wear rubber gloves. When you have finished cleaning off the corrosion, rinse off the battery with warm water. This little bit of "car housekeeping" will minimize the chance of your battery connections' being eaten away by the corroding acids.

Before winter road conditions set in, make sure your car is **tuned up**. As suggested above, a properly tuned engine saves fuel year-round and lasts longer. In addition, make sure your car is ready for winter by checking the following items: ignition system, battery, lights, tire tread, cooling system, fuel system, exhaust system, antifreeze, winter-grade oil, heater, brakes, wiper blades and defroster, snow tires, and chains.

Also, be ready for winter emergencies by making sure you have the following items in your car:

Winter clothing and overshoes	First aid kit
Flashlight and extra batteries	Pocket knife
Transistor radio and extra batteries	Tire chains
Small bag of sand	Shovel
Windshield scraper	Tools

Battery booster cables
Empty three-pound coffee cans
with lids (for burning candles and for sanitation)
A CB radio can also be very
useful in emergencies

Candles and matches
Supply of non-perishable high-
calorie foods

Before starting on a winter vacation, be sure your car is as well prepared as you are. Here is a list of things to check:

- Be sure your heater and defroster are operating properly.
- Check to see if the windshield washer reservoir is full and dispensing the solvent correctly. If wiper blades smear or streak, replace them.
- Have your other fluid levels checked: engine oil, automatic transmission fluid, radiator coolant, battery water and power steering fluid. Also examine the battery cables and hold-down bracket for tightness, the battery case for cracks, radiator cap for water, exhaust system for leaks, and drive belts for proper adjustment.
- Put antifreeze in your car, and attach a tag containing all the pertinent information (dates, quarts, etc.) to the inside of the door. In most situations, winterizing for 40 degrees below zero will take care of any situation.

Warming up the car in winter is another wasteful practice, say the experts, because the car is consuming gasoline needlessly for every minute of idling. Driving slowly for the first couple of miles is a much better way to warm up the car. And, if you pump the accelerator repeatedly before starting, the experts tell us you might be wasting as much as an extra gallon of gasoline a week. Most cars require that you depress the gas pedal just once before starting.

Snow tires and studded tires also increase friction and should be removed as soon as weather permits.

On those snowy days when we must put on the automobile ' lights, even though it is still daylight, here is a little trick to prevent you from leaving those headlights on and running down the battery after parking. Onto the pullout gadget that controls the lights, hang your glove (better a cold hand than a rundown battery that has run out of energy.) Even the key chain out of your purse or similar item attached to the knob will be a reminder to shut off those lights.

Fill several empty milk cartons with **rock salt.** The salt is handy for traction on slippery surfaces, and the milk cartons make good flares when lighted, burning for approximately ten minutes each.

What Car Weight Does to Gas Consumption

According to the Highway User's Federation, every hundred pounds you add to your car (including passengers' weight) costs you between one-tenth and four-tenths miles per gallon. If the car is smaller and lighter, your mileage per gallon will be cut more. One way to cut the load in the summertime is to stop using the trunk of your car to store golf clubs, snow tires from winter, and other unnecessary equipment. In winter, snow and ice on the roof, trunk and hood, and in the fender wells, can add up to more than a hundred pounds.

Miscellaneous Gas Eaters

Short trips are one of the biggest gasoline gulpers, because the engine usually can't warm up to its full efficiency. A short trip can require up to 70 percent more gasoline per mile than a longer trip.

Hard braking, instead of smooth, steady pressure on the pedal, also makes your engine work harder and burn more fuel.

One **fouled spark plug,** misfiring sporadically at 55 m.p.h., can cost you 1.2 miles per gallon of gas. A couple of faulty spark plugs in an eight-cylinder engine can lose five miles per gallon.

A malfunctioning **automatic choke,** which regulates the gasoline-air mixture during starting and warm-up, can cost you 1.5 miles per gallon.

Piling **luggage** on top of the car increases wind drag and uses more gas.

An engine operating under road conditions will warm up faster and lubricate more efficiently than one that is **idling.** Idling, as we have seen, just burns gas (on the average, about one gallon an hour.)

Tailgating and "jack rabbit" starts are murder on gas mileage, since you must constantly brake and accelerate if you let the car in front dictate the speed. On the highway, varying your speed—even by as little as five miles an hour, can reduce fuel economy by as much as 1.3 miles per gallon.

Avoid **unnecessary stops.** It takes up to 20 percent more gas to reach cruising speed from a dead stop than from a speed of just a few miles per hour. So, if the traffic light ahead is red, ease off on the gas and give the light a chance to turn green while you are still moving.

Drive with the **windows closed** whenever possible. Open windows create wind turbulence and have the effect of holding back the car. At highway speeds, open windows can lower mileage as much as 10 percent.

Accessories—Pros and Cons

Cruise-control devices can save gas. For highway driving you can set the control of your desired speed and it automatically takes over the accelerator pedal. A steady speed conserves fuel. (Cruise control is not recommended for mountainous country.)

Other accessories, such as lights, heater, defroster, radio, etc., consume extra fuel and should not be used needlessly.

What about the **automatic transmission?** The experts tell us that in some cases your loss of fuel efficiency with an automatic transmission can be less than one-half mile per gallon; though if you do much driving on city streets, automatic transmission can cost you up to 8 percent more in mileage over a properly shifted manual transmission. However, letting up slightly on the accelerator each time the automatic transmission shifts into a higher gear can decrease this loss a bit. And, if your car has standard transmission, keeping the car in the lower gears too long can be costly. A car running in second gear, for instance, uses about 45 percent more fuel than in fourth gear. **Overdrive units** are now standard equipment or an option on many standard-shift cars. They reduce gas consumption by allowing the engine to work at a fraction of its normal pace at high speeds. Overdrive can improve highway fuel economy by up to 15 percent. (A five-speed transmission serves the same purpose.)

When you flick on that **air-conditioner,** your car's miles per gallon performance immediately drops by about 10 percent. If you do so in heavy stop-and-go traffic, the decrease can fall to 20 percent!

Other **battery-powered accessories**—radio, power windows, tape deck, heater—take extra gas. When used, they activate the alternator-generator to restore to the battery what is being drawn off. The extra energy is paid for with gas burned to provide power to turn the generating unit. Power steering cuts mileage by about one mile per gallon.

Buying a New Car

Be sure to compare **gas mileage.** A luxury sedan can go as few as 8 miles per gallon while a small car with manual transmission can get over 30 miles per gallon. It is estimated that the difference in fuel consumption between a standard American sedan and an economy-sized car is about 260 gallons annually. Before buying, investigate. Check test results in consumer and motor industry magazines. For a free copy of "The Gas Mileage Guide for New Car Buyers," write to: Fuel Economy, Pueblo, Colorado 81009.

Is a smaller, gas-saving car for you? As fuel gets more expensive and more scarce, we see a lot more subcompact cars on the road. (We are defining a subcompact car as having less than one hundred cubic feet of passenger and luggage space, seating for four, and a small, economical engine.) Shop for a subcompact as you would for any other type of car, weighing all important factors such as price, miles per gallons, options, noise, handling and comfort.

Take a **road test** with a subcompact before you buy. You may be surprised at the quick acceleration of many subcompacts. Their lighter weights and efficient engines can make the cars as spirited as V-8-powered cars. And, remember, air-conditioning and automatic transmission will cut acceleration, so be sure to test-drive models with any options you are considering.

Consider the **owner's manual** before buying to determine if the subcompact car is able to tow the camping trailer you own or haul the heavy loads you sometimes have. And don't assume it will be safe to haul or tow weights above specified limits for even short distances. Acceleration will be cut and the engine will likely be strained.

What about the **stability** of a subcompact car in windy weather or when gusts of wind from big trucks hit you? The lower weight of the subcompact is bound to make it slightly more vulnerable than the bigger car, but, because of these cars' excellent directional control, a quick turn of the wheel into the wind when a gust hits will usually put the car back onto a straight course.

Today's front-wheel-drive subcompacts have almost all their weight placed within the wheelbase. This, plus the fact that the front wheels are pulling the car, means **improved traction**. The front-wheel-drive subcompact will usually pull your car through snow, mud, slush and ice that would quite often stop a bigger car. And, most manufacturers contend if you use steel-belted radial tires, you probably won't need snow tires. If you do use snow tires, they should be mounted on the front wheels. Thus, piling extra weight in your little car won't improve—and might very well cut, traction.

Alternatives

Davis, California, has an answer for the auto fuel shortage. They have a **bike path/bike lane system** totalling more than twenty-eight miles! And, there are actually more bicycles in

Davis today than cars—an estimated 30,000 two-wheelers. By using bikes for one-fourth of all trips within the city, Davis residents save roughly 64,000 gallons of gasoline annually and get exercise and fun as well. In addition to promoting travel by bike, this university town (University of California, Davis) operates a seven-route bus system using eighteen coaches, as well as using nine London double-deckers that are five times more efficient than autos, considering the number of people they carry.

Perhaps we could all learn a bit by studying the efforts of this community, which is proclaimed to be the most energy-conscious town in America. (Davis is a small suburban community twelve miles from Sacramento, boasting 36,000 people.)

Overall, **adding people** can save the most on fuel energy of all. Join a car pool so that you drive to work perhaps one day in four, and the savings to each of you will be well worth it. And, since about one-third of all private automobile mileage is for commuting to and from work, joining a car pool makes a lot of sense!

CHAPTER 15

Miscellaneous Energy-Saving Ideas

Landscaping

Think of energy when you landscape. A house protected by a wooden fence or a row of bushes on the windward side may reduce air infiltration inside by up to 30 percent.

Pioneers moving westward were among this nation's first energy conservationists. Settlers heading for the wide-open prairies of Nebraska and the Dakotas actually brought with them— or picked up along the way—tree saplings to plant as **windbreaks** around their farms or ranches.

Today, according to the American Association of Nurserymen (AAN) a South Dakota study was conducted on two identical houses, both exposed to weather but only one protected by trees and shrubs. They found that the tree-protected home required 40 percent less fuel to maintain a constant indoor temperature.

The AAN also recommends a windbreak of **evergreens** to be planted on the north and west sides of the home because that's the direction of most winter winds. The evergreen foliage also has an insulating effect that guards against expensive winter wind, and in summer this same foliage acts as a shield against sunlight, conserving energy. The AAN also recommends planting tall dense **bushes** around the outside air-conditioning unit to improve the efficiency of its performance. These plantings also reduce noise and screen the unit from view. However, do not plant dense bushes too close to hinder unit performance. Leave at least a 2-foot airspace around the units.

It will also help to plant **deciduous trees and vines** on the south, east and west sides of your home. In warm weather they will provide protective shade from the sun, and in cold weather, after the leaves fall, will permit the sun to penetrate and heat those areas.

During Construction

Insulate around and behind all electrical outlets. These areas are often overlooked. Insulating every nook and cranny while the house is being built will keep energy-wasting air convection currents from forming later.

Insulate slab-on-grade floors along the edge of the slab and for two feet horizontally around the edge, under the slab. This cellular foam insulation technique locks out outside air and moisture.

When building a new home or addition, make sure the largest **glass area** faces south. Let the sun help heat your home. Site your house to take advantage of the sun's position in winter and save on heating bills.

Insulate flat or modified vault ceilings. Installing batts, blown-in-place, or rigid foam insulation in the ceiling areas of your home or addition is most easily done during construction.

Insulate between the sill plate and the foundation. Insulation at this junction can help control air infiltration. Then seal the sill with caulk.

Revive the old-fashioned **vestibule.** By creating an air lock area at the entry to your house, a vestibule keeps outdoor drafts away from indoor living areas. If you are replacing an entry door or outfitting the entry to an addition, install an insulated entry door. It will reduce the heat loss/gain through the entry.

Use six-inch studs and ten-inch rafters when framing an addition or a new house. The **wider framing members** allow more space for insulation.

Shade large glass areas with ample **overhangs.** This technique helps control the amount of sun that your windows absorb. Generous overhangs will shade windows and interior spaces from the harsh midday rays and help keep cooling bills under control.

If you are building a new home, consider including a **wall**

mass for thermal storage. Used in conjunction with south-facing glass, interior masonry can absorb the sun's heat during the day, then give off that heat to the living area in the evening.

Other Ideas for Saving Heating and Cooling Energy

If you have an area or room in your home that could be closed off to save heating or cooling, and it is not practical to install a conventional door, a **vinyl accordian door** may be the attractive answer. These units, which cost under $100, are not too difficult to install, and will cover much wider openings than a regulation door. Perhaps one could be used at the bottom of a stairwell leading to bedrooms which are not used during the day.

Electrical companies all over the country advise people to put their window air-conditioners on a **clock-type timer.** However, we have been alerted by the electrical experts that problems could result if you use a timer that is not rated high enough to handle the air-conditioner's load. Smoke will result, the unit will become too hot to handle, etc. So, be sure to purchase a timer "made especially for air-conditioners."

Cover vents in crawl-space foundation walls. These vents are necessary in mild weather to ventilate the crawl-space, but in cold weather they can chill the underside of the floor above. Plastic sheeting, hardboard or exterior plywood can be used to close off these vents. Place the plastic sheeting over the ground if it is not covered.

Some thrifty homemakers put a **clock** in the window so the children who are playing outdoors won't be constantly coming into the warm/or cool house to see what time it is. (Some heating and cooling authorities claim that each child or pet adds about 4 percent to heating and cooling costs just because they have a tendency to go in and out of the house so often.)

Of course, doing the laundry in **cold water** is a great energy saver. However, if you find the powdered detergents are leaving soapy streaks on clothes, instead of dialing to a warmer cycle, mix the detergents with hot tap water in a pitcher, then add it, dissolved, to the machine. No more cold-water problems!

If you have a private or condominium/apartment **swimming pool,** seriously consider *not* heating it. But if your pool is heated, a pool cover can reduce your heating costs by 30 percent . . . and if the covers are removed during the day, additional savings can be chalked up. In many areas, a good pool cover, plus the sun, can effectively eliminate the need for pool heating. A pool temperature of 78 degrees is most desirable, says the American Red Cross, and afternoon swims will avoid costly all-night heating for early morning warm-ups. Since your pool filter pump is your biggest electric user, save by reducing your filter time and by keeping grids and leaf baskets clean. Take maximum advantage of the sun as your heating device.

While You Are Away from Home

Unplug your TV sets. As we have seen, that instant-on feature means your set consumes almost as much electricity when staring blankly into the room as when your family watches.

Your **washer and dryer**—if electric—don't consume electricity when not in use, but they could short out and cause a fire if there is flooding while you are away from home. For safety's sake, unplug them, too. Or turn off the circuit serving these appliances.

If you are going to be gone a few weeks, turn your **water heater** down to the lowest setting or turn it off completely. Electric heaters are shut down at the circuit panel. Gas heaters are best extinguished at the line valve in the gas pipe feeding the heater. (This valve shuts off the pilot light, too.)

During the winter, turn your **furnace** to about 50 degrees. Your pipes won't freeze. In summer, allow the air-conditioner to continue operating, but set the thermostat at about 90 degrees. The house won't stay cool, but the air-conditioner will operate frequently enough to prevent excess humidity to build up that could cause mildew. Allow the circulating fan to operate full time. It won't use much power and will help maintain even humidity.

Unplug that **heater on your water bed,** if you have one.

A unit can consume up to 400 watts—the equivalent of leaving a half-dozen lights on.

Invest in a **timer** to attach to a lamp in the front room. Most people like to leave a light on when they are away to discourage theft and vandalism. But, lights burning all day long not only consume electricity, they also advertise the fact that the homeowners are absent.

Learn to Read Your Electric Meter

Most electric meters have either four or five dials in a row. (The only difference is that the five-dial meter has the capacity to record a higher reading.) Each dial is numbered from 0 to 9, and each has a single rotating hand. As you read the dials from left to right, the numbers indicated by the hand, written down consecutively, show you the figure from which you can determine your usage.

The only way to compute the amount of kilowatt-hours you use in a month is to read your meter for two months in a row. When you subtract your first month's reading from the second one, the difference you get is the total number of kilowatt-hours used during a one-month period. This is the basis on which you are billed.

You should **keep track of your usage.** And, don't be confused if the numbers on your dials are arranged, alternately, clockwise and counterclockwise, as shown in the illustration. This has no bearing on your monthly reading. If any hand points to a space between numbers, the smaller number is the one to note.

Reading Your Gas Bill

Each bill contains several boxes below the customer's address. Reading from the top left you will see a box titled "account number." Each individual has an account number, so check this to determine if you are receiving the proper billing.

To the right of the account number is a box marked, "Current bill for use from . . . to . . . "Under the word "from" should be a date; a different date should appear beneath the word, "to." That will tell you the period of time for which you are being billed.

Appearing next are boxes telling you the number of days of service being billed, the date on which the bill was issued, and the date on which the payment is due.

The box on the far left of the second line is titled, "rate." Beneath the word "rate" there will be either the number 1 or the number 2. Rate 1 indicates general service, and rate 2 indicates a customer with gas space-heating.

Next to the rate box is a box marked "purchased gas charge." In this box there are two numbers—one marked "cents/therm," which indicates the rate the customer is paying per thermal unit of gas. The total amount is marked in the same box under the word "amount."

You will also see a box marked "gas service charge." This will contain another dollars-and-cents figure typed in. This charge represents the gas company's costs such as labor, material and other operating expenses. (Perhaps you had the gas company install a barbeque, or gas fireplace.) Next are boxes marked "municipal tax" and "state tax," which are self-explanatory. The box marked "current bill amount" should show the sum of all the figures beginning with the purchased gas charge.

The third row of boxes represents another breakdown of expenses. The first box gives meter readings, both for the present billing period and the previous one. Subtracting the previous figure from the present figure will give you the amount of gas consumed in 100-cubic-foot figures. Then multiply that figure by the so-called "Btu-factor" to arrive at the total "therms" used.

If you wish to check the billing amount, take the "total therms" number and multiply it by the "cents/therm," and you should arrive at the amount of the purchased gas charge.

Under the current bill amount may be a phrase such as "adj. range," and then a date and a dollar figure. This is an appliance adjustment—charges for special work done at a customer's request on a gas appliance, such as inspection or re-lighting a pilot light on a gas range.

Utilities generally assess a late charge to bills not paid on their due date.

More Money-Saving Ideas

Under a new Government ruling, local utility companies are required to offer "energy audits" free or for a small fee (usually $10 to $15.) The audits are complete evaluations—from basement to roof—that will tell you where the "energy wasters" are in your home and what to do about them.

For summer cooling, consider installing an old-fashioned ceiling fan in your home. It can lower your energy bills dramatically. Ceiling fans, like the ones made popular by the movie "Casablanca," use no more energy than a 15-watt light bulb and can reduce room temperatures in summer by 10 to 12 degrees. They are becoming very popular again today and many decorative styles are being offered. The oldest manufacturer of ceiling fans is The Hunter Fan Company, 2500 Frisco Ave., P.O. Box 14775, Memphis, Tenn. 38114. Many light fixture companies also have them.

Are you wasting your water-bill dollars when you water your lawn? To check how much water your lawn sprinkler is using, place an empty soup can on your grass. An inch of water per week is enough for most lawns. If you are exceeding this amount, reduce the water pressure or cut back on watering.

For free hot water during the summer, leave your garden hose (filled with water) in the sun. When you need a bucket of hot water for a special chore, take it from the hose. The longer the hose, the more hot water it will store.

If you prefer baths, after bathing let the hot water stand in the tub until it cools. The heat from the tub is sufficient to warm and humidify that area of the house on a chilly winter morning or evening.

If you let the hose run while you wash your car, it will cost you dearly. A three-quarter-inch house pours out almost 1,900 gallons per hour—as much water as your family uses in five days. A hose nozzle with a shut-off mechanism will use only a fraction as much water.

Remember, saving energy makes increasing economic sense today. Until 1985, the United States Congress has allowed tax-payers a 15 percent tax credit on the first $2,000 they spend on a variety of energy-saving moves. Among them are insulation, weatherstripping and caulking, storm windows and doors and a number of furnace alterations.

Index

Air-conditioners
  capacity, estimating needs,
    31–33
  cost of operation, 33
  features, 34–35, 164
  maintenance, 36, 37, 51, 129
  thermostat setting, 35, 37
  types, 30
  when you leave home, 165
  See also Solar cooling
Appliances
  energy use by different, 10–11
  operating costs of seventy com-
    mon appliances, 15–18
  owner's manuals, 10
  use of, 11
Automobiles
  accessories, 158–159
  alternatives to driving, 160–161
  car-pooling, 161
  car-washing, 168
  diesel, 149–150
  energy use of, in U.S., 140
  gasohol, 147–149, 157
  gasoline, 142–147
  motor oil, 150–151
  new car buying, 159–160
  tips for better mileage,
    141–142, 157–158
  tires, 152–154
  tune-ups, 140–141
  winter car care, 154–157

Bathrooms, heat from bath-water,
  168

Caulking. *See* Weatherstripping
Chimneys, 130
  See also under Wood heat
Clothes washers, -dryers
  dryers, 23–24
  operation, 22, 165
  selection and features, 22–23
  water levels, 23
  water temperatures, 21–22, 164
Coffee makers (electric), 11
Cooking. *See* Stoves
Cooling houses, tips for, 36–38
  See also Air-conditioners

Degree days, 50
Dehumidifiers, 97–98
Dishes, hand washing, 7, 8
Dishwashers
  operation, 8
  selection, 7
  water savings with, 7
Doors, 121
Draperies, insulated, 120

Electric blankets, 12–14
Electric heaters, 53
Electric meters, 166

171